More Praise for the Fantasy Sports and Mathematics Series

"'This is the greatest assignment ever!' Hearing that from one of my students was more than enough to know that I had a great way to get them to enjoy learning mathematics. Thanks for putting this great program together, we LOVE IT!"

—Jeff Thompson, Walt Morey Middle School,
Troutdale, Oregon

"The Fantasy Sports and Mathematics programs allow you to reach the students, making the lessons not only comprehensive and enriching, but very exciting. The students loved it, worked harder than ever, and their GEPA scores increased over 40 points in the first year alone!"

—Robert Creamer, Woodbine School District,
Woodbine, New Jersey

"Fantasy Football and Mathematics is a great resource! My students have never been so excited about a unit. I even have former students participating in it on their own!"

—Jason Williams, Azalea Middle School,
St. Petersburg, Florida

"The beauty of the Fantasy Sports and Mathematics programs is that while all the students are busy having fun, enjoying the competition, and building better relationships with sports-minded parents, they are, much to their surprise, learning to love math."

—Ryan D. Verver, Southwest Chicago Christian School,
Tinley Park, Illinois

"Playing Fantasy Football and Mathematics from a preservice teacher perspective allows my secondary math methods students to see the power of teaching math with motivational and mathematically sound materials. My students are very engaged in fantasy football."

—Dana Pomykal Franz, assistant professor,
curriculum and instruction, Mississippi State University

"Students were eager to be the first in their math class to tally their scores and begin working with their team's stats. It was great to see boys and girls who were working both above and below grade level fully engaged in using their math skills and learning new ones."

—Sara Suchman, Harvard Graduate School of Education,
former middle school director

Jossey-Bass Teacher

Jossey-Bass Teacher provides K–12 teachers with essential knowledge and tools to create a positive and lifelong impact on student learning. Trusted and experienced educational mentors offer practical classroom-tested and theory-based teaching resources for improving teaching practice in a broad range of grade levels and subject areas. From one educator to another, we want to be your first source to make every day your best day in teaching. *Jossey-Bass Teacher* resources serve two types of informational needs—essential knowledge and essential tools.

Essential knowledge resources provide the foundation, strategies, and methods from which teachers may design curriculum and instruction to challenge and excite their students. Connecting theory to practice, essential knowledge books rely on a solid research base and time-tested methods, offering the best ideas and guidance from many of the most experienced and well-respected experts in the field.

Essential tools save teachers time and effort by offering proven, ready-to-use materials for in-class use. Our publications include activities, assessments, exercises, instruments, games, ready reference, and more. They enhance an entire course of study, a weekly lesson, or a daily plan. These essential tools provide insightful, practical, and comprehensive materials on topics that matter most to K–12 teachers.

Fantasy Sports and Mathematics Series

Fantasy Baseball and Mathematics: A Resource Guide for Teachers and Parents, Grades 5 and Up

Fantasy Baseball and Mathematics: Student Workbook

Fantasy Basketball and Mathematics: A Resource Guide for Teachers and Parents, Grades 5 and Up

Fantasy Basketball and Mathematics: Student Workbook

Fantasy Football and Mathematics: A Resource Guide for Teachers and Parents, Grades 5 and Up

Fantasy Football and Mathematics: Student Workbook

Fantasy Soccer and Mathematics: A Resource Guide for Teachers and Parents, Grades 5 and Up

Fantasy Soccer and Mathematics: Student Workbook

Fantasy Baseball and Mathematics

A Resource Guide for Teachers and Parents

Dan Flockhart

JOSSEY-BASS

Published by Jossey-Bass
A Wiley Imprint
989 Market Street, San Francisco, CA 94103-1741   www.josseybass.com

Jossey-Bass books and products are available through most bookstores. To contact Jossey-Bass directly, call our Customer Care Department within the U.S. at 800-956-7739, outside the U.S. at 317-572-3986, or fax 317-572-4002.

Jossey-Bass also publishes its books in a variety of electronic formats. Some content that appears in print may not be available in electronic books.

ISBN: 978-0-7879-9443-3

Printed in the United States of America
FIRST EDITION
PB Printing   10 9 8 7 6 5 4 3 2 1

About This Book: FAQs

What is Fantasy Baseball and Mathematics?
Fantasy Baseball and Mathematics is a game that is played by millions of adolescents and adults nationwide. Participants create fantasy teams by selecting professional baseball players. The baseball players earn points based on their performances in their games. Each week, students use newspapers or online resources to locate players' statistics in order to find out how many total points were earned by their team. The object of the game is to accumulate the highest number of points.

When is the best time of the year to play Fantasy Baseball and Mathematics?
Major League Baseball begins in early April and ends in late September. Most teachers will play the game from April until the end of the school year. However, teachers who have the same students in consecutive years can begin in April and continue the game when school begins in the fall. Fantasy Baseball and Mathematics is also a perfect fit for summer school programs.

How much time is required to play the game?
Students can compute their weekly points in 15 to 40 minutes.

Where can I find the players' statistics?
Players' statistics are found in box scores (summaries of games) in newspapers, as well as online at www.fantasysportsmath.com and at other sports Web sites.

How much preparation time is required for teachers?
Once the game begins, little preparation time is required if you have access to online resources or newspapers-in-education programs.

Where can I find help?
A blog, podcast, and support forum for teachers can be found at www.fantasysportsmath.com. Teachers can enter their students in contests on that site as well.

How do I assess the skills that are covered?
This book contains forty-six practice worksheets and forty-six corresponding assessments in the form of quizzes. In addition, a comprehensive pretest/posttest is included.

About the Author

Dan Flockhart received his multiple-subject teaching credential from California State University, East Bay in 1988. He taught mathematics in grades 5 through 8 for eleven years at St. Matthew's Episcopal Day School in San Mateo, California, where he incorporated fantasy sports into his math curriculum. He has also taught general studies classes at College of the Redwoods in Eureka, California. He received a master of arts degree in education from Humboldt State University in 2005; the title of his thesis was "Teacher Perceptions of the Effects of Fantasy Football in the Teaching of Mathematics." Flockhart has enjoyed participating in fantasy sports for over twenty-five years.

In addition to authoring the Fantasy Sports and Mathematics series, Flockhart maintains a Web site, www.fantasysportsmath.com, where teachers can participate in forums and contests and find out more about the series.

Acknowledgments

This book would not have been possible without the help and support of all of the teachers who provided me with feedback. In addition, I am grateful to Tiffany for her continual support as well as the many hours she spent on this project. I am also thankful to Kate, who made all of this possible. You are one of my angels! I was also lucky to work with wonderful production editors, Elizabeth and Susan, and copyeditors, Carolyn and Bev. They were fun to work with and I was impressed with their willingness to do whatever it took to produce the best possible product. My thanks go out as well to Chris for creating one of the best covers I've ever seen. In addition, I'm grateful to Lena for ensuring that all the math is accurate. Finally, I'd like to thank Annie for being by my side every step of the way.

Contents

Part One: Computing Weekly Points   1

Chapter One: How to Play Fantasy Baseball and Mathematics   3

Chapter Two: Explaining Fantasy Baseball and Mathematics to Students: Handouts

Chapter Five: Quizzes

To Dad

Introduction

Welcome to Fantasy Baseball and Mathematics, an exciting and easy-to-use program that takes advantage of the fantasy sports phenomenon. I created this educational program by combining twenty-five years of participation in fantasy sports and eleven years of experience teaching mathematics in grades 5 through 8. This curriculum was the most successful one I used in the classroom; both boys and girls ran into my room on a daily basis, wanting to play the game.

Fantasy Baseball and Mathematics is a game in which students select and manage teams of professional, college, or high school baseball players. The baseball players earn points based on their performances in real games. Each week, students use newspapers or online resources to locate their players' statistics in order to find out how many points were earned by their team. The goal of the game is to accumulate the highest number of points.

This comprehensive resource guide is supported by research (see www.fantasysportsmath.com). The program includes lesson plans, student handouts, graphing activities, a pretest/posttest, forty-six practice worksheets, forty-six quizzes, and more than one hundred scoring systems that give teachers and parents the flexibility to customize the content according to the skill level of the students. Fantasy Baseball and Mathematics addresses the social and cognitive needs of students because it is based on a pedagogy that is student-centered. Students work collaboratively in groups to compute their weekly points, check their peers' answers, and create their graphs. Students become active learners as they select their teams and starting lineups. Consequently, this autonomy helps students build their decision-making skills. Students also receive multiple exposures to concepts, facilitating mastery. Moreover, the hands-on features (technology, newspapers, and graphing activities) address all three learning styles: auditory (learning by listening), visual (learning by seeing), and kinesthetic or tactile (learning by doing and touching). These features help to meet the needs of all students and result in a high level of student interest.

Fantasy Baseball and Mathematics has three components. The first element is computation of weekly points. The second element consists of graphing activities, and is optional. The third component is integration of practice worksheets and corresponding quizzes into the game, which is also optional.

All players' names in this text are fabricated. However, students will select real baseball players. Consequently, the game is fun and dynamic. In addition, preparation time for teachers is minimal once the game has begun.

Teachers using this program can enter their students in contests at www.fantasysportsmath.com. A support forum for teachers can be found there as well. Let's get started!

National Council of Teachers of Mathematics Standards and Expectations Addressed by *Fantasy Baseball and Mathematics*

Portions of the standards that appear in italics are not covered.

Number and Operations Standard for Grades 3–5
Understand numbers, ways of representing numbers, relationships among numbers, and number systems:
- Understand the place-value structure of the base-ten number system and be able to represent and compare whole numbers and decimals
- Develop understanding of fractions as parts of unit wholes, as parts of a collection, as locations on number lines, and as divisions of whole numbers
- Recognize and generate equivalent forms of commonly used fractions, decimals, and percents
- Explore numbers less than 0 by extending the number line and through familiar applications

Understand meanings of operations and how they relate to one another:
- Understand the effects of multiplying and dividing whole numbers
- Understand and use properties of operations, such as the distributivity of multiplication over addition

Compute fluently and make reasonable estimates:
- Develop fluency in adding, subtracting, multiplying, and dividing whole numbers

Algebra Standard for Grades 3–5
Understand patterns, relations, and functions:
- Represent and analyze patterns and functions, using words, tables, and graphs

Represent and analyze mathematical situations and structures, using algebraic symbols:
- Represent and analyze mathematical situations and structures, using algebraic symbols
- Identify such properties as commutativity, associativity, and distributivity and use them to compute with whole numbers
- Represent the idea of a variable as an unknown quantity, using a letter or a symbol
- Express mathematical relationships, using equations

Measurement Standard for Grades 3–5
- Understand such attributes as length, area, weight, *volume, and size of angle* and select the appropriate type of unit for measuring each attribute
- Carry out simple unit conversions, such as from centimeters to meters, within a system of measurement
- Select and apply appropriate standard units and tools to measure length, area, *volume,* weight, *time, temperature, and the size of angles*
- Develop, understand, and use formulas to find the area of rectangles *and related triangles and parallelograms*

Data Analysis and Probability Standard for Grades 3–5
- Represent data, using tables and graphs such as *line plots,* bar graphs, and line graphs

Select and use appropriate statistical methods to analyze data:
- Propose and justify conclusions and predictions that are based on data *and design studies to further investigate the conclusions or predictions*

Understand and apply basic concepts of probability:
- Predict the probability of outcomes of simple experiments and test the predictions
- Understand that the measure of the likelihood of an event can be represented by a number from 0 to 1

Problem-Solving Standard for Grades Prekindergarten–12

- Build new mathematical knowledge through problem solving
- Solve problems that arise in mathematics and in other contexts
- Apply and adapt a variety of appropriate strategies to solve problems
- Monitor and reflect on the process of mathematical problem solving

Reasoning and Proof Standard for Grades Prekindergarten–12

- Recognize reasoning and proof as fundamental aspects of mathematics
- Make and investigate mathematical conjectures

Communication Standard for Grades Prekindergarten–12

- Organize and consolidate mathematical thinking through communication
- Communicate mathematical thinking coherently and clearly to peers, teachers, and others
- Use the language of mathematics to express mathematical ideas precisely

Connections Standard for Grades Prekindergarten–12

- Recognize and use connections among mathematical ideas
- Understand how mathematical ideas interconnect and build on one another to produce a coherent whole
- Recognize and apply mathematics in contexts outside of mathematics

Representation Standard for Grades Prekindergarten–12

- Create and use representations to organize, record, and communicate mathematical ideas
- Select, apply, and translate among mathematical representations to solve problems

Number and Operations Standard for Grades 6–8

Understand numbers, ways of representing numbers, relationships among numbers, and number systems:

- Work flexibly with fractions, decimals, and percents to solve problems
- Compare and order fractions, decimals, and percents efficiently and find their approximate locations on a number line
- Understand and use ratios and proportions to represent quantitative relationships
- Develop an understanding of large numbers and recognize and appropriately use exponential, scientific, and *calculator* notation
- Use factors, multiples, prime factorization, *and relatively prime numbers* to solve problems
- Develop meaning for integers and represent and compare quantities with them

Understand meanings of operations and how they relate to one another:

- Understand the meaning and effects of arithmetic operations with fractions, decimals, and integers
- Use the associative and commutative properties of addition and multiplication and the distributive property of multiplication over addition to simplify computations with integers, fractions, and decimals
- Understand and use the inverse relationships of addition and subtraction, multiplication and division, and squaring and finding square roots to simplify computations and solve problems

Compute fluently and make reasonable estimates:

- *Develop* and analyze algorithms for computing with fractions, decimals, and integers and develop fluency in their use

Algebra Standard for Grades 6–8
Understand patterns, relations, and functions:

- Develop an initial conceptual understanding of different uses of variables
- Recognize and generate equivalent forms for simple algebraic expressions and solve linear equations

Measurement Standard for Grades 6–8
Understand measurable attributes of objects and the units, systems, and processes of measurement:

- Understand both metric and customary systems of measurement
- Understand relationships among units and convert from one unit to another within the same system

Apply appropriate techniques, tools, and formulas to determine measurements:

- Select and apply techniques and tools to accurately find length, area, *volume,* and angle measures to appropriate levels of precision
- Develop and use formulas to determine the circumference of circles and the *area of triangles, parallelograms, trapezoids,* and circles and *develop strategies to find the area of more-complex shapes*
- Solve problems involving scale factors, using ratio and proportion

Data Analysis and Probability Standard for Grades 6–8

- Select, create, and use appropriate graphical representations of data, including histograms, *box plots,* and scatter plots

Number and Operations Standard for Grades 9–12

- Develop an understanding of permutations and combinations as counting techniques

Algebra Standard for Grades 9–12

Understand patterns, relations, and functions:

- Understand *relations* and functions *and select, convert flexibly among, and use various representations for them*

Represent and analyze mathematical situations and structures, using algebraic symbols:

- Use symbolic algebra to represent and explain mathematical relationships

Data Analysis and Probability Standard for Grades 9–12

- Understand histograms, *parallel box plots,* and scatter plots and use them to display data

Computing Weekly Points

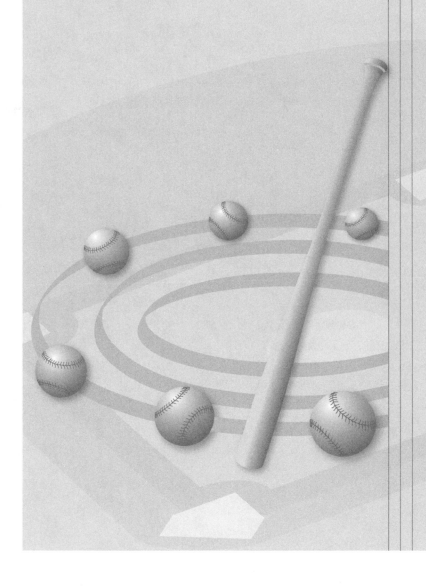

Chapter One

How to Play Fantasy Baseball and Mathematics

Fantasy Baseball and Mathematics is a game in which participants create and manage teams of professional baseball players. Players earn points for hits, walks, stolen bases, home runs, runs scored, and runs batted in. Players lose points for striking out or making errors. Each week, students find the sum of the points earned by their players, using one of the scoring systems in this book. The object of the game is to accumulate the highest number of points.

> **How to Play the Game**
> Step 1: Selecting players
> Step 2: Reading box scores
> Step 3: Collecting data
> Step 4: Computing points

Step 1: Selecting Players

There are two options for selecting players. Option 1 includes a salary cap and player values. Player values and salary caps will be updated before each season and posted at www.fantasysportsmath.com. The process of creating new player values is time-consuming and requires research and extensive knowledge of players' performance over the last several years. Additional factors taken into account when assigning player values include current injuries, whether a player recently changed teams, and so on. The purchase of this book entitles you to one season of free player values. Lists of player values for subsequent seasons will be provided for a nominal fee. To access player values, visit www.fantasysportsmath.com. Click on "Player Values" and follow the instructions. Your password is w2m4c2c8. This password can be used one time only, at which time it will expire.

In option 2, you avoid the salary cap and player values, but students do not receive several benefits of these critical components of the game, which are explained later.

Option 1: Permanent Teams with Salary Cap

Students have a salary cap of $40 million. This is the amount they can spend on player values. Students select thirteen players at the positions listed in Table 1.1. Selecting pitchers is optional because they have different scoring criteria from everyday players. For younger students, it may be easier to not select pitchers. If students do not select pitchers, the salary cap is $35 million.

Table 1.1 lists the number of players to be selected at each position as well as the number of players in a starting lineup for each position.

Table 1.1. Baseball Positions: Number to Be Selected and Number in a Starting Lineup

Position	Number to Be Selected	Number in a Starting Lineup
First base (1B)	1	1
Second base (2B)	1	1
Third base (3B)	1	1
Shortstop (SS)	1	1
Catcher (C)	1	1
Outfield (OF)	4	3
Designated hitter (DH)	1	1
Pitcher (P)	2	1
Infield (IF)	1	

How to play Fantasy Baseball and Mathematics

Notice that students will select three substitute players (one infielder, one outfielder, and one pitcher) to use when a starting player gets injured or is performing poorly. More than one student can select the same player.

You can choose whether students have to set their starting lineups at the end of each week for the following week or if they can use the statistics from the best performing players for each week. For example, a student could compute the points for all four of her outfielders, then select the three outfielders who earned the most points that week. I used the first method because it was less time-consuming.

The main advantage of using option 1 is that it promotes equality in the game. If students spend close to the cap on their players' values, the quality of their teams should be relatively equal. I had many students (both girls and boys) who knew very little about baseball yet managed to do very well and, in several instances, even win the game. Another advantage is that students have to compromise as they select players because the salary cap is structured so that they cannot simply select the top players at each position. This allows them to hone their decision-making skills, which facilitates their cognitive development. Students can also make trades.

Another benefit of using option 1 is that students get to work with large numbers as they attempt to spend as close as possible to the salary cap. Moreover, in addition to circle graphs, students will be able to construct stacked-bar and multi-line graphs to track player performance over time because they will use the same players for the duration of the game. Finally, if a player is determined to be out for the year, students can use the portion of the salary cap they spent on that player to purchase another player.

For all of the preceding reasons, option 1 is recommended.

Option 2: Different Teams Each Week

Each week, each student selects one team. For example, a student may decide to select her hometown team for the first week of the game. However, she will not be allowed to choose that team in later weeks because each team can be selected only once by each student during the game. Unlike in option 1, students compute points using team statistics rather than statistics of individual players. For example, if a team had a total of sixteen hits in a game, that is the number that would be used to compute points.

If you use option 1 to select players, students' rosters will remain the same for the length of the game, unless there are trades. If you use option 2, students' teams will change from week to week. *Note that the handouts, graphs, and worksheets in this text are all based on option 1.*

Choosing Your Own Team

In addition to having students choose players, you should also create your own fantasy team. You can use your team as an example and to help assess students' work. Students also enjoy competing with their teachers or parents.

Trades

Students may trade players if they selected players using option 1. For example, a student may want to trade a second baseman for an outfielder. Consequently, that student would insert her substitute infielder into the starting lineup. When a trade is consummated, it is important that the students involved in the trade make the necessary alterations to their fantasy team roster. Teachers may want to limit the number of trades to five or ten per student. Salary cap numbers do not apply to trades.

Injuries

If you cannot locate a player's name in the box scores, he is probably injured or taking a day off. *If this occurs, the players' score is counted as zero.* Likewise, if a student's pitcher does not pitch in a given week, the pitcher's score is counted as zero. If a player is declared out for the year and if students used option 1 to select players, a student can use the portion of the salary cap that was spent on that player to purchase another player. A list of injured players can be found in newspapers as well as online at www.fantasysportsmath.com or on other sports Web sites.

Step 2: Reading Box Scores

Let's look at the statistics for Sammy Cooke, the designated hitter for the Rats. In Table 1.2, in the line next to his name, you can see that he scored one run and had three hits, five runs batted in, one base on balls, and zero strikeouts. At the bottom of the box score, notice that he also hit a home run.

Table 1.2.   Sample Box Score

Bats	ab	r	h	rbi	bb	so	lob	avg
N Colt ss	5	0	0	0	0	1	2	.293
T Flyer cf	5	1	2	0	0	1	0	.260
F Vargas 1b	5	0	2	1	0	1	1	.385
J Macky rf	5	2	3	0	0	0	1	.287
A Cortez 3b	4	2	2	0	1	0	1	.298
T Joon dh	3	0	1	1	1	0	4	.318
C Flores lf	4	1	1	2	0	0	2	.257
G Hollis c	3	0	1	2	0	1	2	.161
a–M Vilipane ph-c	1	0	0	0	0	0	0	.293

(Cont'd.)

Table 1.2. Sample Box Score (*Cont'd.*)

Bats	ab	r	h	rbi	bb	so	lob	avg
M Roper 2b	3	0	0	0	1	0	1	.111
J Ezatz Jr pr-2b	0	0	0	0	0	0	0	.269
Totals	38	6	12	6	3	4	14	

BATTING: 2B—T Flyer (9, C Targe); J Macky (15, P Garcia). 3B—T Flyer (1, C Targe).
BASERUNNING: **SB**—C Flores (3, 2nd base off C Targe/J Blanco).
FIELDING: **E**—M Roper (1, ground ball).

Rats	ab	r	h	rbi	bb	so	lob	avg
L Carter ss	4	1	1	0	1	1	4	.299
R Renault 2b	5	1	1	0	0	0	3	.275
R Brady rf	3	3	2	0	2	1	2	.300
J Martinez 3b	4	2	3	2	1	1	1	.328
S Cooke dh	4	1	3	5	1	0	1	.286
R Davis pr-dh	0	0	0	0	0	0	0	.308
J Blanco c	4	1	2	1	1	0	2	.297
J Smith 1b	3	0	0	0	0	0	4	.242
T Allen 1b	1	0	0	0	1	0	1	.232
B Johnson cf	4	0	2	1	1	1	4	.238
T Blake lf	4	0	1	0	0	0	2	.244
Totals	36	9	15	9	8	4	24	

BATTING: 2B—J Martinez (13, J Carrillo); R Brady (17, J Carrillo); S Cooke (19, L Deringer). **HR**—S Cooke (7, 7th inning off W Brown 1 on, 1 out).

Note: ab = at bats; **r** = **runs; h** = **hits; rbi** = **runs batted in; bb** = **bases on balls** (i.e., **walk); so** = **strikeouts;** lob = left on base; avg = average. Batting: 2B = double; 3B = triple; **HR** = **home run;** S = sacrifice. Baserunning: SB = stolen base. Items in bold will be used in the Fantasy Baseball and Math game. Fielding: **E** = **error.**

Table 1.3. Sample Box Score: Pitchers' Statistics

Bats	ip	h	r	er	bb	so	hr	era
J Carrillo	$6\frac{1}{3}$	9	6	6	6	3	0	3.56
W Brown (L, 2–1;)	0	2	1	1	1	0	1	2.57
M Reddmon	$\frac{2}{3}$	3	2	2	1	1	0	3.73
L Deringer	1	1	0	0	0	0	0	4.82

Rats	ip	h	r	er	bb	so	hr	era
C Targe	$5\frac{1}{3}$	10	6	6	1	3	0	4.53
P Garcia	$1\frac{1}{3}$	1	0	0	1	0	0	6.75
T Estrada (W, 1–1)	$\frac{2}{3}$	0	0	0	0	0	0	6.57
T Lloyd (H, 13)	$\frac{2}{3}$	0	0	0	1	1	0	2.64
S Wolly (S, 15)	1	1	0	0	0	0	0	1.05

Note: ip = **innings pitched**; h = **hits**; r = **runs**; er = earned runs; **bb** = **bases on balls (i.e., walk)**; so = **strikeouts**; hr = home runs; era = earned run average. L = loss; W = win; H = number of holds (i.e., the number of times a pitcher has entered a game in a save situation and left the game with his team leading); S = save.

Now let's look at statistics for pitchers. In Table 1.3, notice that Juan Carrillo pitched $6\frac{1}{3}$ innings. Carrillo gave up nine hits, six runs, six bases on balls, and he had three strikeouts.

Additional statistics you will follow include errors and stolen bases for everyday players, as well as wins by a pitcher. When a pitcher wins a game, you will see a "W" next to his name in the box score. Notice that the winning pitcher for this game was T. Estrada of the Rats.

Players who have stolen bases or committed errors will be listed in the "Baserunning" or "Fielding" sections. For example, notice that Carlos Flores stole a base for the Bats (his third of the year; if he had stolen two bases, it would have been listed like this: C Flores 2 (3,4 2nd base off C Targe/J Blanco; 3rd base off C Targe/J Blanco). Also notice that Matt Roper committed one error for the Bats.

Step 3: Collecting Data

Each week, students use newspapers or online resources to access data from one game in which each of the players in their starting lineup participated.

How to play Fantasy Baseball and Mathematics

Students can choose the game that produced the best statistics for each player. Options for collecting data include the following:

1. Enroll your class in a newspapers-in-education program in order to receive free copies of newspapers.

2. If it is not possible to enroll in a newspapers-in-education program, choose a couple of students to cut box scores out of one newspaper and make copies for the other students. Students can reference the baseball standings in the newspaper to ensure that they have cut out at least one box score for each team. This duty can be rotated.

3. Have students visit www.fantasysportsmath.com, and do the following:

   a. Click the "Get Baseball Stats" link.

   b. On the following page, use the calendar to select any day from the previous week.

   c. Find a team one of your players participated in and click on the box score for that game. Students can find the game during the previous week in which each of their players produced the best statistics.

If students use online resources to collect data, they can choose from a number of games their players participated in for the previous week because

Table 1.4. Default Scoring System for Nonpitchers

For Each:	Players Earn:		
Home run (HR)	$\frac{1}{2}$	or	.500
Run scored (R)	$\frac{1}{3}$	or	.333
Run batted in (RBI)	$\frac{1}{3}$	or	.333
Hit (H)	$\frac{1}{6}$	or	.167
Stolen base (SB)	$\frac{1}{7}$	or	.143
Base on balls (BB)	$\frac{1}{7}$	or	.143
Strikeout (SO)	$-\frac{1}{21}$	or	$-.048$
Error (E)	$-\frac{1}{21}$	or	$-.048$

Table 1.5. Default Scoring System for Pitchers

For Each:	Players Earn:		
Win (W)	$\dfrac{1}{2}$	or	.500
Inning pitched (IP)*	$\dfrac{1}{3}$	or	.333
Strikeout (SO)	$\dfrac{1}{6}$	or	.167
Run allowed (R)	$-\dfrac{1}{7}$	or	−.143
Hit allowed (H)	$-\dfrac{1}{21}$	or	−.048
Base on balls allowed (BB)	$-\dfrac{1}{21}$	or	−.048

Note: All decimals are rounded to the nearest thousandth.

*Rounded down to the nearest whole number—for example, $6\frac{2}{3}$ would be rounded down to 6.

statistics are archived online. Students can also access data if they missed a week or two. Using online resources is the quickest and easiest method.

Step 4: Computing Points

Tables 1.4 and 1.5 show the default scoring systems for nonpitchers and pitchers.

To keep things simple for younger students, you have the option of not selecting pitchers. Since they pitch only once every five days or so, in many cases, their names will not be in box scores anyway.

Before we learn how to compute points, we need a team. A hypothetical starting team is listed in Table 1.6. All players on this team are from the previous Bats-Rats box score (Tables 1.2 and 1.3). Normally, all players will not be found in the same box score because students usually select players from several teams. Three players from this team (Felipe Vargas, Louie Carter, and Julio Martinez) are also on the fantasy team called the English Bulldogs and are used extensively as examples throughout this book. All graphs and several worksheets are linked to these three players.

The points earned by players can be computed by two different methods. One method uses algebra; the other method does not. If students use both methods to compute points, they can verify their results. However, if students do not have the skills to work with variables in linear equations, they can use the nonalgebraic method to compute points.

Table 1.7 (on page 13) shows the computation of points earned for the English Bulldogs using the non-algebraic method with the default scoring system. I recommend using this method for the first few weeks and then introducing

Table 1.6. The English Bulldogs

Felipe Vargas	First base (1B)
Matt Roper	Second base (2B)
Julio Martinez	Third base (3B)
Louie Carter	Shortstop (SS)
Jose Blanco	Catcher (C)
Carlos Flores	Outfield (OF)
Tim Flyer	Outfield (OF)
Ray Brady	Outfield (OF)
Sammy Cooke	Designated hitter (DH)
Juan Carrillo	Pitcher (P)

students to the algebraic method. The default scoring system can be used each week to determine the ranking of students' teams in the game. It was designed so that students can plot the weekly points earned for their players to precise numerical values on stacked-bar and multiple-line graphs. This is explained later. However, if you wish, you may choose a different scoring system to meet the skill level of your students.

Once students have mastered the non-algebraic method of computing points, they can move on to the algebraic method, which includes the use of linear equations that contain variables. These equations are known as *total points equations*. Younger students may be initially intimidated by the algebraic look of the equations. However, once they have used them a few times, they become comfortable with the equations and feel proud that they are doing algebra. The default total points equations (the algebraic method) and the default scoring system (the non-algebraic method) contain the same numerical values. Consequently, students can check their work if they use both methods because both methods will result in the same answer. The default total points equations (one for everyday players and one for pitchers) are listed in the next section.

Default Total Points Equation for Nonpitchers

Numerical values are the same as in the default scoring system.

$$\frac{1}{2}(H) + \frac{1}{3}(R + I) + \frac{1}{6}(B) + \frac{1}{7}(S + W) - \frac{1}{21}(K + E) = T$$

H = number of home runs
R = number of runs scored
I = number of runs batted in
B = number of hits
S = number of stolen bases
W = number of bases on balls (walks)
K = number of strikeouts
E = number of errors
T = total points earned for one week for one player

Default Total Points Equation for Pitchers

$$\frac{1}{2}(V) + \frac{1}{3}(P) + \frac{1}{6}(K) - \frac{1}{7}(R) - \frac{1}{21}(B + W) = T$$

V = number of wins
P = number of innings pitched, rounded down to the nearest whole number
K = number of strikeouts
R = number of runs allowed
B = number of hits allowed
W = number of bases on balls allowed
T = total points earned for one week for one pitcher

Note: If you think your students will be confused by using two linear equations, then don't select pitchers.

How to play Fantasy Baseball and Mathematics

Table 1.7. Points Earned for the English Bulldogs

Nonpitchers

Player	Number of Home Runs $\times \frac{1}{2}$	Number of Runs Scored and RBIs $\times \frac{1}{3}$	Number of Hits $\times \frac{1}{6}$	Number of Stolen Bases and Bases on Balls $\times \frac{1}{7}$	Number of Strikeouts and Errors $\times \left(-\frac{1}{21}\right)$	Total Individual Points
Vargas	0	$\frac{1}{3}$	$\frac{2}{6}$	0	$-\frac{1}{21}$	$\frac{13}{21}$
Roper	0	0	0	$\frac{1}{7}$	$-\frac{1}{21}$	$\frac{2}{21}$
Martinez	0	$\frac{4}{3}$	$\frac{3}{6}$	$\frac{1}{7}$	$-\frac{1}{21}$	$1\frac{13}{14}$
Carter	0	$\frac{1}{3}$	$\frac{1}{6}$	$\frac{1}{7}$	$-\frac{1}{21}$	$\frac{25}{42}$
Blanco	0	$\frac{2}{3}$	$\frac{2}{6}$	$\frac{1}{7}$	0	$1\frac{1}{7}$
Flores	0	$\frac{3}{3}$	$\frac{1}{6}$	$\frac{1}{7}$	0	$1\frac{13}{42}$
Flyer	0	$\frac{1}{3}$	$\frac{2}{6}$	0	$-\frac{1}{21}$	$\frac{13}{21}$
Brady	0	$\frac{3}{3}$	$\frac{2}{6}$	$\frac{2}{7}$	$-\frac{1}{21}$	$1\frac{4}{7}$
Cooke	$\frac{1}{2}$	$\frac{6}{3}$	$\frac{3}{6}$	$\frac{1}{7}$	0	$3\frac{1}{7}$

Pitcher

Player	Number of Wins $\times \frac{1}{2}$	Number of Innings Pitched $\times \frac{1}{3}$	Number of Strikeouts $\times \frac{1}{6}$	Number of Runs Allowed $\times \left(-\frac{1}{7}\right)$	Number of Hits and Bases on Balls Allowed $\times \left(-\frac{1}{21}\right)$	Total Individual Points
Carrillo	0	$\frac{6}{3}$	$\frac{3}{6}$	$-\frac{6}{7}$	$-\frac{15}{21}$	$\frac{13}{14}$

Total team points: $\frac{13}{21} + \frac{2}{21} + 1\frac{13}{14} + \frac{25}{42} + 1\frac{1}{7} + 1\frac{13}{42} + \frac{13}{21} + 1\frac{4}{7} + 3\frac{1}{7} + \frac{13}{14} = 11\frac{40}{42} = 11\frac{20}{21}$

Example Using Default Total Points Equations: Points Earned for the English Bulldogs

$$\frac{1}{2}(H) + \frac{1}{3}(R + I) + \frac{1}{6}(B) + \frac{1}{7}(S + W) - \frac{1}{21}(K + E) = T$$

Felipe Vargas

$$\frac{1}{2}(0) + \frac{1}{3}(0 + 1) + \frac{1}{6}(2) + \frac{1}{7}(0 + 0) - \frac{1}{21}(1 + 0) = \frac{13}{21}$$

Matt Roper

$$\frac{1}{2}(0) + \frac{1}{3}(0 + 0) + \frac{1}{6}(0) + \frac{1}{7}(0 + 1) - \frac{1}{21}(0 + 1) = \frac{2}{21}$$

Julio Martinez

$$\frac{1}{2}(0) + \frac{1}{3}(2 + 2) + \frac{1}{6}(3) + \frac{1}{7}(0 + 1) - \frac{1}{21}(1 + 0) = 1\frac{13}{14}$$

Louie Carter

$$\frac{1}{2}(0) + \frac{1}{3}(1 + 0) + \frac{1}{6}(1) + \frac{1}{7}(0 + 1) - \frac{1}{21}(1 + 0) = \frac{25}{42}$$

Jose Blanco

$$\frac{1}{2}(0) + \frac{1}{3}(1 + 1) + \frac{1}{6}(2) + \frac{1}{7}(0 + 1) - \frac{1}{21}(0 + 0) = 1\frac{1}{7}$$

Carlos Flores

$$\frac{1}{2}(0) + \frac{1}{3}(1 + 2) + \frac{1}{6}(1) + \frac{1}{7}(1 + 0) - \frac{1}{21}(0 + 0) = 1\frac{13}{42}$$

Tim Flyer

$$\frac{1}{2}(0) + \frac{1}{3}(1 + 0) + \frac{1}{6}(2) + \frac{1}{7}(0 + 0) - \frac{1}{21}(1 + 0) = \frac{13}{21}$$

Ray Brady

$$\frac{1}{2}(0) + \frac{1}{3}(3 + 0) + \frac{1}{6}(2) + \frac{1}{7}(0 + 2) - \frac{1}{21}(1 + 0) = 1\frac{4}{7}$$

Sammy Cooke

$$\frac{1}{2}(1) + \frac{1}{3}(1 + 5) + \frac{1}{6}(3) + \frac{1}{7}(0 + 1) - \frac{1}{21}(0 + 0) = 3\frac{1}{7}$$

For Juan Carrillo, use the default total points equation for pitchers:

$$\frac{1}{2}(V) + \frac{1}{3}(P) + \frac{1}{6}(K) - \frac{1}{7}(R) - \frac{1}{21}(B + W) = T$$

Juan Carrillo

$$\frac{1}{2}(0) + \frac{1}{3}(6) + \frac{1}{6}(3) - \frac{1}{7}(6) - \frac{1}{21}(9 + 6) = \frac{13}{14}$$

Total points:

$$7\frac{208}{42} = 11\frac{40}{42} = 11\frac{20}{21}$$

Notice that both methods of computing points earned resulted in the same answer. Similarly, students will be able to check their work by using both methods.

Additional Scoring Systems

Over 160 scoring systems are listed in the following pages. You can choose a scoring system that is appropriate for the skill level of your students. The different scoring systems give students opportunities to work with roots, exponents, summations, factorials, integers, fractions, decimals, and absolute value. For my students, I used one scoring system throughout the game and supplemented that system with additional scoring systems when students were ready. Students should use one scoring system throughout the game to determine their cumulative points and to update their stacked-bar and multiple-line graphs.

Many scoring systems are more advanced than the default scoring system, especially those that are based on negative numerical values. In those systems, the goal is to acquire the least amount of points (or the greatest absolute value). Acquiring the least amount of points is an effective way to teach the concept of absolute value. In the following example, a pitcher might earn $\frac{3}{4}$ if a student used scoring system number four. However, if the student placed absolute value symbols around scoring system number five before using it to compute points, the player would also earn $\frac{3}{4}$.

4. $\quad \frac{1}{8}(1) + \frac{1}{12}(7) + \frac{1}{16}(5) - \frac{1}{24}(2) - \frac{1}{48}(6 + 3) = \frac{3}{4}$

5. $\quad \left| -\frac{1}{8}(1) - \frac{1}{12}(7) - \frac{1}{16}(5) + \frac{1}{24}(2) + \frac{1}{48}(6 + 3) \right| = \frac{3}{4}$

Consequently, students can check their work by using both positive and negative versions of the same scoring systems, since both will result in the same absolute value. In order to do this, you can simply insert absolute value symbols around any scoring systems that are based on negative numerical values. It is possible (though unlikely) that a baseball player may earn a negative amount of points even if students are using equations that are based on positive numerical values. In other words, a player may have a bad game statistically and not generate enough positive points to offset his negative points. You can prevent this scenario by using equation number one (which doesn't use any negative values) or by informing students that the lowest score for a player for one week will be zero, thus ensuring that younger students will not be confused by negative numerical values.

Scoring systems are categorized according to their content and whether or not they use relative proportionality (that is, whether points earned have equal ratios). For example, scoring systems numbers four and six (located below) use relative proportionality because the ratios between the fractions in each equation are the same. In other words, a home run (H) in the first equation is 1.5 times greater than a run scored or run batted in $(R$ and $I)$, two times greater than a hit (B), three times greater than a stolen base or walk $(S$ and $W)$, and so on. Likewise, these ratios are the same between the numerical values in the second equation.

4. $\quad \frac{1}{8}(H) + \frac{1}{12}(R + I) + \frac{1}{16}(B) + \frac{1}{24}(S + W) - \frac{1}{48}(K + E) = T$

6. $\quad \frac{1}{6}(H) + \frac{1}{9}(R + I) + \frac{1}{12}(B) + \frac{1}{18}(S + W) - \frac{1}{36}(K + E) = T$

The advantage of using scoring systems that use relative proportionality is that you can use these different scoring systems during the course of the game without unfairly changing the rankings. In other words, a student whose team earned the highest number of points in a given week will earn the highest number of points in that week no matter which scoring systems are used, so long as the scoring systems are proportionate. Conversely, let's say you used a scoring system that was based on fractions for the first ten weeks, and then used a different scoring system for week 11 that was based on factorials and not proportionate to the original scoring system you used. It is possible that the student who was in last place after ten weeks could leap into first place after week 11 if her team performed strongly, because the scoring systems based on factorials are not proportionate and can result in teams earning hundreds of points in one week. Consequently, it's not fair for a student who has built up a small lead over the course of ten weeks to suddenly be hundreds of points out of the lead based on one week. For this reason, I suggest using the same scoring system or scoring systems that are proportionate throughout the game in order to determine standings. If you wish to include other scoring systems, I would not include these to determine the rankings of the students' teams.

Scoring systems 4–35, 42–71, and 72–97 use relative proportionality.

In each scoring system, the first equation is for nonpitchers and the second equation is for pitchers. If you do not select pitchers, then you will use only the first equation in each set.

Additional Scoring Systems (Total Points Equations)

The first equation in each set is for nonpitchers; the second equation is for pitchers.

Integers

1. $5 (H) + 4 (R + I) + 3 (B) + 2 (S + W) = T$
   $5 (V) + 4 (P) + 3 (K) = T$

2. $5 (H) + 4 (R + I) + 3 (B) + 2 (S + W) - 1 (K + E) = T$
   $5 (V) + 4 (P) + 3 (K) - 2 (R) - 1 (B + W) = T$

3. $-5 (H) - 4 (R + I) - 3 (B) - 2 (S + W) + 1 (K + E) = T$
   $-5 (V) - 4 (P) - 3 (K) + 2 (R) + 1 (B + W) = T$

Fractions

Equations 4–35 use relative proportionality.

4. $\frac{1}{8} (H) + \frac{1}{12} (R + I) + \frac{1}{16} (B) + \frac{1}{24} (S + W) - \frac{1}{48} (K + E) = T$

   $\frac{1}{8} (V) + \frac{1}{12} (P) + \frac{1}{16} (K) - \frac{1}{24} (R) - \frac{1}{48} (B + W) = T$

5. $-\frac{1}{8} (H) - \frac{1}{12} (R + I) - \frac{1}{16} (B) - \frac{1}{24} (S + W) + \frac{1}{48} (K + E) = T$

   $-\frac{1}{8} (V) - \frac{1}{12} (P) - \frac{1}{16} (K) + \frac{1}{24} (R) + \frac{1}{48} (B + W) = T$

6. $\frac{1}{6} (H) + \frac{1}{9} (R + I) + \frac{1}{12} (B) + \frac{1}{18} (S + W) - \frac{1}{36} (K + E) = T$

   $\frac{1}{6} (V) + \frac{1}{9} (P) + \frac{1}{12} (K) - \frac{1}{18} (R) - \frac{1}{36} (B + W) = T$

7. $-\frac{1}{6} (H) - \frac{1}{9} (R + I) - \frac{1}{12} (B) - \frac{1}{18} (S + W) + \frac{1}{36} (K + E) = T$

   $-\frac{1}{6} (V) - \frac{1}{9} (P) - \frac{1}{12} (K) + \frac{1}{18} (R) + \frac{1}{36} (B + W) = T$

8. $\frac{1}{9} (H) + \frac{1}{13.5} (R + I) + \frac{1}{18} (B) + \frac{1}{27} (S + W) - \frac{1}{54} (K + E) = T$

   $\frac{1}{9} (V) + \frac{1}{13.5} (P) + \frac{1}{18} (K) - \frac{1}{27} (R) - \frac{1}{54} (B + W) = T$

9. $-\frac{1}{9} (H) - \frac{1}{13.5} (R + I) - \frac{1}{18} (B) - \frac{1}{27} (S + W) + \frac{1}{54} (K + E) = T$

   $-\frac{1}{9} (V) - \frac{1}{13.5} (P) - \frac{1}{18} (K) + \frac{1}{27} (R) + \frac{1}{54} (B + W) = T$

10. $\dfrac{1}{10}(H) + \dfrac{1}{15}(R + I) + \dfrac{1}{20}(B) + \dfrac{1}{30}(S + W) - \dfrac{1}{60}(K + E) = T$

$\dfrac{1}{10}(V) + \dfrac{1}{15}(P) + \dfrac{1}{20}(K) - \dfrac{1}{30}(R) - \dfrac{1}{60}(B + W) = T$

11. $-\dfrac{1}{10}(H) - \dfrac{1}{15}(R + I) - \dfrac{1}{20}(B) - \dfrac{1}{30}(S + W) + \dfrac{1}{60}(K + E) = T$

$-\dfrac{1}{10}(V) - \dfrac{1}{15}(P) - \dfrac{1}{20}(K) + \dfrac{1}{30}(R) + \dfrac{1}{60}(B + W) = T$

12. $\dfrac{1}{12}(H) + \dfrac{1}{18}(R + I) + \dfrac{1}{24}(B) + \dfrac{1}{36}(S + W) - \dfrac{1}{72}(K + E) = T$

$\dfrac{1}{12}(V) + \dfrac{1}{18}(P) + \dfrac{1}{24}(K) - \dfrac{1}{36}(R) - \dfrac{1}{72}(B + W) = T$

13. $-\dfrac{1}{12}(H) - \dfrac{1}{18}(R + I) - \dfrac{1}{24}(B) - \dfrac{1}{36}(S + W) + \dfrac{1}{72}(K + E) = T$

$-\dfrac{1}{12}(V) - \dfrac{1}{18}(P) - \dfrac{1}{24}(K) + \dfrac{1}{36}(R) + \dfrac{1}{72}(B + W) = T$

14. $\dfrac{1}{4}(H) + \dfrac{1}{6}(R + I) + \dfrac{1}{8}(B) + \dfrac{1}{12}(S + W) - \dfrac{1}{24}(K + E) = T$

$\dfrac{1}{4}(V) + \dfrac{1}{6}(P) + \dfrac{1}{8}(K) - \dfrac{1}{12}(R) - \dfrac{1}{24}(B + W) = T$

15. $-\dfrac{1}{4}(H) - \dfrac{1}{6}(R + I) - \dfrac{1}{8}(B) - \dfrac{1}{12}(S + W) + \dfrac{1}{24}(K + E) = T$

$-\dfrac{1}{4}(V) - \dfrac{1}{6}(P) - \dfrac{1}{8}(K) + \dfrac{1}{12}(R) + \dfrac{1}{24}(B + W) = T$

16. $\dfrac{1}{2}(H) + \dfrac{1}{3}(R + I) + \dfrac{1}{4}(B) + \dfrac{1}{6}(S + W) - \dfrac{1}{12}(K + E) = T$

$\dfrac{1}{2}(V) + \dfrac{1}{3}(P) + \dfrac{1}{4}(K) - \dfrac{1}{6}(R) - \dfrac{1}{12}(B + W) = T$

17. $-\dfrac{1}{2}(H) - \dfrac{1}{3}(R + I) - \dfrac{1}{4}(B) - \dfrac{1}{6}(S + W) + \dfrac{1}{12}(K + E) = T$

$-\dfrac{1}{2}(V) - \dfrac{1}{3}(P) - \dfrac{1}{4}(K) + \dfrac{1}{6}(R) + \dfrac{1}{12}(B + W) = T$

18. $\dfrac{1}{14}(H) + \dfrac{1}{21}(R + I) + \dfrac{1}{28}(B) + \dfrac{1}{42}(S + W) - \dfrac{1}{84}(K + E) = T$

$\dfrac{1}{14}(V) + \dfrac{1}{21}(P) + \dfrac{1}{28}(K) - \dfrac{1}{42}(R) - \dfrac{1}{84}(B + W) = T$

19. $-\dfrac{1}{14}(H) - \dfrac{1}{21}(R + I) - \dfrac{1}{28}(B) - \dfrac{1}{42}(S + W) + \dfrac{1}{84}(K + E) = T$

$-\dfrac{1}{14}(V) - \dfrac{1}{21}(P) - \dfrac{1}{28}(K) + \dfrac{1}{42}(R) + \dfrac{1}{84}(B + W) = T$

20. $\dfrac{1}{16}(H) + \dfrac{1}{24}(R + I) + \dfrac{1}{32}(B) + \dfrac{1}{48}(S + W) - \dfrac{1}{96}(K + E) = T$

$\dfrac{1}{16}(V) + \dfrac{1}{24}(P) + \dfrac{1}{32}(K) - \dfrac{1}{48}(R) - \dfrac{1}{96}(B + W) = T$

21. $-\dfrac{1}{16}(H) - \dfrac{1}{24}(R + I) - \dfrac{1}{32}(B) - \dfrac{1}{48}(S + W) + \dfrac{1}{96}(K + E) = T$

$-\dfrac{1}{16}(V) - \dfrac{1}{24}(P) - \dfrac{1}{32}(K) + \dfrac{1}{48}(R) + \dfrac{1}{96}(B + W) = T$

22. $\dfrac{1}{18}(H) + \dfrac{1}{27}(R + I) + \dfrac{1}{36}(B) + \dfrac{1}{54}(S + W) - \dfrac{1}{108}(K + E) = T$

$\dfrac{1}{18}(V) + \dfrac{1}{27}(P) + \dfrac{1}{36}(K) - \dfrac{1}{54}(R) - \dfrac{1}{108}(B + W) = T$

23. $-\dfrac{1}{18}(H) - \dfrac{1}{27}(R + I) - \dfrac{1}{36}(B) - \dfrac{1}{54}(S + W) + \dfrac{1}{108}(K + E) = T$

$-\dfrac{1}{18}(V) - \dfrac{1}{27}(P) - \dfrac{1}{36}(K) + \dfrac{1}{54}(R) + \dfrac{1}{108}(B + W) = T$

24. $\dfrac{1}{1.\overline{6}}(H) + \dfrac{1}{2.5}(R + I) + \dfrac{1}{3.\overline{3}}(B) + \dfrac{1}{5}(S + W) - \dfrac{1}{10}(K + E) = T$

$\dfrac{1}{1.\overline{6}}(V) + \dfrac{1}{2.5}(P) + \dfrac{1}{3.\overline{3}}(K) - \dfrac{1}{5}(R) - \dfrac{1}{10}(B + W) = T$

25. $-\dfrac{1}{1.\overline{6}}(H) - \dfrac{1}{2.5}(R + I) - \dfrac{1}{3.\overline{3}}(B) - \dfrac{1}{5}(S + W) + \dfrac{1}{10}(K + E) = T$

$-\dfrac{1}{1.\overline{6}}(V) - \dfrac{1}{2.5}(P) - \dfrac{1}{3.\overline{3}}(K) + \dfrac{1}{5}(R) + \dfrac{1}{10}(B + W) = T$

26. $\dfrac{1}{83.\overline{3}}(H) + \dfrac{1}{125}(R + I) + \dfrac{1}{166.\overline{6}}(B) + \dfrac{1}{250}(S + W) - \dfrac{1}{500}(K + E) = T$

$\dfrac{1}{83.\overline{3}}(V) + \dfrac{1}{125}(P) + \dfrac{1}{166.\overline{6}}(K) - \dfrac{1}{250}(R) - \dfrac{1}{500}(B + W) = T$

27. $-\dfrac{1}{83.\overline{3}}(H) - \dfrac{1}{125}(R + I) - \dfrac{1}{166.\overline{6}}(B) - \dfrac{1}{250}(S + W) + \dfrac{1}{500}(K + E) = T$

$-\dfrac{1}{83.\overline{3}}(V) - \dfrac{1}{125}(P) - \dfrac{1}{166.\overline{6}}(K) + \dfrac{1}{250}(R) + \dfrac{1}{500}(B + W) = T$

28. $\dfrac{1}{50}(H) + \dfrac{1}{75}(R + I) + \dfrac{1}{100}(B) + \dfrac{1}{150}(S + W) - \dfrac{1}{300}(K + E) = T$

$\dfrac{1}{50}(V) + \dfrac{1}{75}(P) + \dfrac{1}{100}(K) - \dfrac{1}{150}(R) - \dfrac{1}{300}(B + W) = T$

29. $-\dfrac{1}{50}(H) - \dfrac{1}{75}(R + I) - \dfrac{1}{100}(B) - \dfrac{1}{150}(S + W) + \dfrac{1}{300}(K + E) = T$

$-\dfrac{1}{50}(V) - \dfrac{1}{75}(P) - \dfrac{1}{100}(K) + \dfrac{1}{150}(R) + \dfrac{1}{300}(B + W) = T$

30. $\dfrac{1}{166.\overline{6}}\,(H) + \dfrac{1}{250}\,(R+I) + \dfrac{1}{333.\overline{3}}\,(B) + \dfrac{1}{500}\,(S+W) - \dfrac{1}{1000}\,(K+E) = T$

$\dfrac{1}{166.\overline{6}}\,(V) + \dfrac{1}{250}\,(P) + \dfrac{1}{333.\overline{3}}\,(K) - \dfrac{1}{500}\,(R) - \dfrac{1}{1000}\,(B+W) = T$

31. $-\dfrac{1}{166.\overline{6}}\,(H) - \dfrac{1}{250}\,(R+I) - \dfrac{1}{333.\overline{3}}\,(B) - \dfrac{1}{500}\,(S+W) + \dfrac{1}{1000}\,(K+E) = T$

$-\dfrac{1}{166.\overline{6}}\,(V) - \dfrac{1}{250}\,(P) - \dfrac{1}{333.\overline{3}}\,(K) + \dfrac{1}{500}\,(R) + \dfrac{1}{1000}\,(B+W) = T$

32. $\dfrac{1}{13.5}\,(H) + \dfrac{1}{20.25}\,(R+I) + \dfrac{1}{27}\,(B) + \dfrac{1}{40.5}\,(S+W) - \dfrac{1}{81}\,(K+E) = T$

$\dfrac{1}{13.5}\,(V) + \dfrac{1}{20.25}\,(P) + \dfrac{1}{27}\,(K) - \dfrac{1}{40.5}\,(R) - \dfrac{1}{81}\,(B+W) = T$

33. $-\dfrac{1}{13.5}\,(H) - \dfrac{1}{20.25}\,(R+I) - \dfrac{1}{27}\,(B) - \dfrac{1}{40.5}\,(S+W) + \dfrac{1}{81}\,(K+E) = T$

$-\dfrac{1}{13.5}\,(V) - \dfrac{1}{20.25}\,(P) - \dfrac{1}{27}\,(K) + \dfrac{1}{40.5}\,(R) + \dfrac{1}{81}\,(B+W) = T$

34. $\dfrac{1}{16.\overline{6}}\,(H) + \dfrac{1}{25}\,(R+I) + \dfrac{1}{33.\overline{3}}\,(B) + \dfrac{1}{50}\,(S+W) - \dfrac{1}{100}\,(K+E) = T$

$\dfrac{1}{16.\overline{6}}\,(V) + \dfrac{1}{25}\,(P) + \dfrac{1}{33.\overline{3}}\,(K) - \dfrac{1}{50}\,(R) - \dfrac{1}{100}\,(B+W) = T$

35. $-\dfrac{1}{16.\overline{6}}\,(H) - \dfrac{1}{25}\,(R+I) - \dfrac{1}{33.\overline{3}}\,(B) - \dfrac{1}{50}\,(S+W) + \dfrac{1}{100}\,(K+E) = T$

$-\dfrac{1}{16.\overline{6}}\,(V) - \dfrac{1}{25}\,(P) - \dfrac{1}{33.\overline{3}}\,(K) + \dfrac{1}{50}\,(R) + \dfrac{1}{100}\,(B+W) = T$

36. $\dfrac{5}{6}\,(H) + \dfrac{4}{5}\,(R+I) + \dfrac{3}{4}\,(B) + \dfrac{2}{7}\,(S+W) - \dfrac{2}{8}\,(K+E) = T$

$\dfrac{5}{6}\,(V) + \dfrac{4}{5}\,(P) + \dfrac{3}{4}\,(K) - \dfrac{2}{7}\,(R) - \dfrac{2}{8}\,(B+W) = T$

37. $-\dfrac{5}{6}\,(H) - \dfrac{4}{5}\,(R+I) - \dfrac{3}{4}\,(B) - \dfrac{2}{7}\,(S+W) + \dfrac{2}{8}\,(K+E) = T$

$-\dfrac{5}{6}\,(V) - \dfrac{4}{5}\,(P) - \dfrac{3}{4}\,(K) + \dfrac{2}{7}\,(R) + \dfrac{2}{8}\,(B+W) = T$

38. $\dfrac{1}{2}\,(H) + \dfrac{1}{3}\,(R+I) + \dfrac{1}{4}\,(B) + \dfrac{1}{5}\,(S+W) - \dfrac{1}{6}\,(K+E) = T$

$\dfrac{1}{2}\,(V) + \dfrac{1}{3}\,(P) + \dfrac{1}{4}\,(K) - \dfrac{1}{5}\,(R) - \dfrac{1}{6}\,(B+W) = T$

39. $-\dfrac{1}{2}\,(H) - \dfrac{1}{3}\,(R+I) - \dfrac{1}{4}\,(B) - \dfrac{1}{5}\,(S+W) + \dfrac{1}{6}\,(K+E) = T$

$-\dfrac{1}{2}\,(V) - \dfrac{1}{3}\,(P) - \dfrac{1}{4}\,(K) + \dfrac{1}{5}\,(R) + \dfrac{1}{6}\,(B+W) = T$

How to play Fantasy Baseball and Mathematics

40. $\frac{1}{2}(H) + \frac{1}{4}(R + I) + \frac{1}{8}(B) + \frac{1}{16}(S + W) - \frac{1}{32}(K + E) = T$

$\frac{1}{2}(V) + \frac{1}{4}(P) + \frac{1}{8}(K) - \frac{1}{16}(R) - \frac{1}{32}(B + W) = T$

41. $-\frac{1}{2}(H) - \frac{1}{4}(R + I) - \frac{1}{8}(B) - \frac{1}{16}(S + W) + \frac{1}{32}(K + E) = T$

$-\frac{1}{2}(V) - \frac{1}{4}(P) - \frac{1}{8}(K) + \frac{1}{16}(R) + \frac{1}{32}(B + W) = T$

Decimals

Equations 42–71 use relative proportionality.

42. $.6(H) + .4(R + I) + .3(B) + .2(S + W) - .1(K + E) = T$
$.6(V) + .4(P) + .3(K) - .2(R) - .1(B + W) = T$

43. $-.6(H) - .4(R + I) - .3(B) - .2(S + W) + .1(K + E) = T$
$-.6(V) - .4(P) - .3(K) + .2(R) + .1(B + W) = T$

44. $1.2(H) + .8(R + I) + .6(B) + .4(S + W) - .2(K + E) = T$
$1.2(V) + .8(P) + .6(K) - .4(R) - .2(B + W) = T$

45. $-1.2(H) - .8(R + I) - .6(B) - .4(S + W) + .2(K + E) = T$
$-1.2(V) - .8(P) - .6(K) + .4(R) + .2(B + W) = T$

46. $1.8(H) + 1.2(R + I) + .9(B) + .6(S + W) - .3(K + E) = T$
$1.8(V) + 1.2(P) + .9(K) - .6(R) - .3(B + W) = T$

47. $-1.8(H) - 1.2(R + I) - .9(B) - .6(S + W) + .3(K + E) = T$
$-1.8(V) - 1.2(P) - .9(K) + .6(R) + .3(B + W) = T$

48. $2.4(H) + 1.6(R + I) + 1.2(B) + .8(S + W) - .4(K + E) = T$
$2.4(V) + 1.6(P) + 1.2(K) - .8(R) - .4(B + W) = T$

49. $-2.4(H) - 1.6(R + I) - 1.2(B) - .8(S + W) + .4(K + E) = T$
$-2.4(V) - 1.6(P) - 1.2(K) + .8(R) + .4(B + W) = T$

50. $3.0(H) + 2.0(R + I) + 1.5(B) + 1.0(S + W) - .5(K + E) = T$
$3.0(V) + 2.0(P) + 1.5(K) - 1.0(R) - .5(B + W) = T$

51. $-3.0(H) - 2.0(R + I) - 1.5(B) - 1.0(S + W) + .5(K + E) = T$
$-3.0(V) - 2.0(P) - 1.5(K) + 1.0(R) + .5(B + W) = T$

52. $3.6(H) + 2.4(R + I) + 1.8(B) + 1.2(S + W) - .6(K + E) = T$
$3.6(V) + 2.4(P) + 1.8(K) - 1.2(R) - .6(B + W) = T$

53. $-3.6(H) - 2.4(R + I) - 1.8(B) - 1.2(S + W) + .6(K + E) = T$
$-3.6(V) - 2.4(P) - 1.8(K) + 1.2(R) + .6(B + W) = T$

54. $4.2(H) + 2.8(R + I) + 2.1(B) + 1.4(S + W) - .7(K + E) = T$
$4.2(V) + 2.8(P) + 2.1(K) - 1.4(R) - .7(B + W) = T$

55. $-4.2(H) - 2.8(R + I) - 2.1(B) - 1.4(S + W) + .7(K + E) = T$
$-4.2(V) - 2.8(P) - 2.1(K) + 1.4(R) + .7(B + W) = T$

56. $4.8 \, (H) + 3.2 \, (R + I) + 2.4 \, (B) + 1.6 \, (S + W) - .8 \, (K + E) = T$

    $4.8 \, (V) + 3.2 \, (P) + 2.4 \, (K) - 1.6 \, (R) - .8 \, (B + W) = T$

57. $-4.8 \, (H) - 3.2 \, (R + I) - 2.4 \, (B) - 1.6 \, (S + W) + .8 \, (K + E) = T$

    $-4.8 \, (V) - 3.2 \, (P) - 2.4 \, (K) + 1.6 \, (R) + .8 \, (B + W) = T$

58. $5.4 \, (H) + 3.6 \, (R + I) + 2.7 \, (B) + 1.8 \, (S + W) - .9 \, (K + E) = T$

    $5.4 \, (V) + 3.6 \, (P) + 2.7 \, (K) - 1.8 \, (R) - .9 \, (B + W) = T$

59. $-5.4 \, (H) - 3.6 \, (R + I) - 2.7 \, (B) - 1.8 \, (S + W) + .9 \, (K + E) = T$

    $-5.4 \, (V) - 3.6 \, (P) - 2.7 \, (K) + 1.8 \, (R) + .9 \, (B + W) = T$

60. $.06 \, (H) + .04 \, (R + I) + .03 \, (B) + .02 \, (S + W) - .01 \, (K + E) = T$

    $.06 \, (V) + .04 \, (P) + .03 \, (K) - .02 \, (R) - .01 \, (B + W) = T$

61. $-.06 \, (H) - .04 \, (R + I) - .03 \, (B) - .02 \, (S + W) + .01 \, (K + E) = T$

    $-.06 \, (V) - .04 \, (P) - .03 \, (K) + .02 \, (R) + .01 \, (B + W) = T$

62. $.006 \, (H) + .004 \, (R + I) + .003 \, (B) + .002 \, (S + W) - .001 \, (K + E) = T$

    $.006 \, (V) + .004 \, (P) + .003 \, (K) - .002 \, (R) - .001 \, (B + W) = T$

63. $-.006 \, (H) - .004 \, (R + I) - .003 \, (B) - .002 \, (S + W) + .001 \, (K + E) = T$

    $-.006 \, (V) - .004 \, (P) - .003 \, (K) + .002 \, (R) + .001 \, (B + W) = T$

64. $.30 \, (H) + .20 \, (R + I) + .15 \, (B) + .10 \, (S + W) - .05 \, (K + E) = T$

    $.30 \, (V) + .20 \, (P) + .15 \, (K) - .10 \, (R) - .05 \, (B + W) = T$

65. $-.30 \, (H) - .20 \, (R + I) - .15 \, (B) - .10 \, (S + W) + .05 \, (K + E) = T$

    $-.30 \, (V) - .20 \, (P) - .15 \, (K) + .10 \, (R) + .05 \, (B + W) = T$

66. $.03 \, (H) + .02 \, (R + I) + .015 \, (B) + .01 \, (S + W) - .005 \, (K + E) = T$

    $.03 \, (V) + .02 \, (P) + .015 \, (K) - .01 \, (R) - .005 \, (B + W) = T$

67. $-.03 \, (H) - .02 \, (R + I) - .015 \, (B) - .01 \, (S + W) + .005 \, (K + E) = T$

    $-.03 \, (V) - .02 \, (P) - .015 \, (K) + .01 \, (R) + .005 \, (B + W) = T$

68. $1.5 \, (H) + 1 \, (R + I) + .75 \, (B) + .5 \, (S + W) - .25 \, (K + E) = T$

    $1.5 \, (V) + 1 \, (P) + .75 \, (K) - .5 \, (R) - .25 \, (B + W) = T$

69. $-1.5 \, (H) - 1 \, (R + I) - .75 \, (B) - .5 \, (S + W) + .25 \, (K + E) = T$

    $-1.5 \, (V) - 1 \, (P) - .75 \, (K) + .5 \, (R) + .25 \, (B + W) = T$

70. $.15 \, (H) + .1 \, (R + I) + .075 \, (B) + .05 \, (S + W) - .025 \, (K + E) = T$

    $.15 \, (V) + .1 \, (P) + .075 \, (K) - .05 \, (R) - .025 \, (B + W) = T$

71. $-.15 \, (H) - .1 \, (R + I) - .075 \, (B) - .05 \, (S + W) + .025 \, (K + E) = T$

    $-.15 \, (V) - .1 \, (P) - .075 \, (K) + .05 \, (R) + .025 \, (B + W) = T$

Fractions and Decimals

Equations 72–97 use relative proportionality.

72. $.25 \, (H) + \dfrac{1}{6} \, (R + I) + .125 \, (B) + \dfrac{1}{12} \, (S + W) - \dfrac{1}{24} \, (K + E) = T$

    $.25 \, (V) + \dfrac{1}{6} \, (P) + .125 \, (K) - \dfrac{1}{12} \, (R) - \dfrac{1}{24} \, (B + W) = T$

        How to play Fantasy Baseball and Mathematics

73. $-.25\ (H) - \dfrac{1}{6}\ (R + I) - .125\ (B) - \dfrac{1}{12}\ (S + W) + \dfrac{1}{24}\ (K + E) = T$

   $-.25\ (V) - \dfrac{1}{6}\ (P) - .125\ (K) + \dfrac{1}{12}\ (R) + \dfrac{1}{24}\ (B + W) = T$

74. $.2\ (H) + \dfrac{1}{7.5}\ (R + I) + .1\ (B) + \dfrac{1}{15}\ (S + W) - \dfrac{1}{30}\ (K + E) = T$

   $.2\ (V) + \dfrac{1}{7.5}\ (P) + .1\ (K) - \dfrac{1}{15}\ (R) - \dfrac{1}{30}\ (B + W) = T$

75. $-.2\ (H) - \dfrac{1}{7.5}\ (R + I) - .1\ (B) - \dfrac{1}{15}\ (S + W) + \dfrac{1}{30}\ (K + E) = T$

   $-.2\ (V) - \dfrac{1}{7.5}\ (P) - .1\ (K) + \dfrac{1}{15}\ (R) + \dfrac{1}{30}\ (B + W) = T$

76. $.5\ (H) + .1\ (R + I) + \dfrac{1}{15}\ (B) + \dfrac{1}{30}\ (S + W) - \dfrac{1}{60}\ (K + E) = T$

   $.5\ (V) + .1\ (P) + \dfrac{1}{15}\ (K) - \dfrac{1}{30}\ (R) - \dfrac{1}{60}\ (B + W) = T$

77. $-.5\ (H) - .1\ (R + I) - \dfrac{1}{15}\ (B) - \dfrac{1}{30}\ (S + W) + \dfrac{1}{60}\ (K + E) = T$

   $-.5\ (V) - .1\ (P) - \dfrac{1}{15}\ (K) + \dfrac{1}{30}\ (R) + \dfrac{1}{60}\ (B + W) = T$

78. $.04\ (H) + \dfrac{1}{37.5}\ (R + I) + .02\ (B) + \dfrac{1}{75}\ (S + W) - \dfrac{1}{150}\ (K + E) = T$

   $.04\ (V) + \dfrac{1}{37.5}\ (P) + .02\ (K) - \dfrac{1}{75}\ (R) - \dfrac{1}{150}\ (B + W) = T$

79. $-.04\ (H) - \dfrac{1}{37.5}\ (R + I) - .02\ (B) - \dfrac{1}{75}\ (S + W) + \dfrac{1}{150}\ (K + E) = T$

   $-.04\ (V) - \dfrac{1}{37.5}\ (P) - .02\ (K) + \dfrac{1}{75}\ (R) + \dfrac{1}{150}\ (B + W) = T$

80. $.01\ (H) + \dfrac{1}{150}\ (R + I) + .005\ (B) + \dfrac{1}{300}\ (S + W) - \dfrac{1}{600}\ (K + E) = T$

   $.01\ (V) + \dfrac{1}{150}\ (P) + .005\ (K) - \dfrac{1}{300}\ (R) - \dfrac{1}{600}\ (B + W) = T$

81. $-.01\ (H) - \dfrac{1}{150}\ (R + I) - .005\ (B) - \dfrac{1}{300}\ (S + W) + \dfrac{1}{600}\ (K + E) = T$

   $-.01\ (V) - \dfrac{1}{150}\ (P) - .005\ (K) + \dfrac{1}{300}\ (R) + \dfrac{1}{600}\ (B + W) = T$

82. $\dfrac{3}{500}\ (H) + .004\ (R + I) + \dfrac{3}{1000}\ (B) + .002\ (S + W) - \dfrac{1}{1000}\ (K + E) = T$

   $\dfrac{3}{500}\ (V) + .004\ (P) + \dfrac{3}{1000}\ (K) - .002\ (R) - \dfrac{1}{1000}\ (B + W) = T$

83. $-\dfrac{3}{500}\,(H) - .004\,(R + I) - \dfrac{3}{1000}\,(B) - .002\,(S + W) + \dfrac{1}{1000}\,(K + E) = T$

$-\dfrac{3}{500}\,(V) - .004\,(P) - \dfrac{3}{1000}\,(K) + .002\,(R) + \dfrac{1}{1000}\,(B + W) = T$

84. $\dfrac{3}{40}\,(H) + \dfrac{1}{20}\,(R + I) + .0375\,(B) + .025\,(S + W) - .0125\,(K + E) = T$

$\dfrac{3}{40}\,(V) + \dfrac{1}{20}\,(P) + .0375\,(K) - .025\,(R) - .0125\,(B + W) = T$

85. $-\dfrac{3}{40}\,(H) - \dfrac{1}{20}\,(R + I) - .0375\,(B) - .025\,(S + W) + .0125\,(K + E) = T$

$-\dfrac{3}{40}\,(V) - \dfrac{1}{20}\,(P) - .0375\,(K) + .025\,(R) + .0125\,(B + W) = T$

86. $.06\,(H) + \dfrac{1}{25}\,(R + I) + .03\,(B) + \dfrac{1}{50}\,(S + W) - .01\,(K + E) = T$

$.06\,(V) + \dfrac{1}{25}\,(P) + .03\,(K) - \dfrac{1}{50}\,(R) - .01\,(B + W) = T$

87. $-.06\,(H) - \dfrac{1}{25}\,(R + I) - .03\,(B) - \dfrac{1}{50}\,(S + W) + .01\,(K + E) = T$

$-.06\,(V) - \dfrac{1}{25}\,(P) - .03\,(K) + \dfrac{1}{50}\,(R) + .01\,(B + W) = T$

88. $\dfrac{1}{10}\,(H) + .0\overline{6}\,(R + I) + \dfrac{1}{20}\,(B) + .0\overline{3}\,(S + W) - .01\overline{6}\,(K + E) = T$

$\dfrac{1}{10}\,(V) + .0\overline{6}\,(P) + \dfrac{1}{20}\,(K) - .0\overline{3}\,(R) - .01\overline{6}\,(B + W) = T$

89. $-\dfrac{1}{10}\,(H) - .0\overline{6}\,(R + I) - \dfrac{1}{20}\,(B) - .0\overline{3}\,(S + W) + .01\overline{6}\,(K + E) = T$

$-\dfrac{1}{10}\,(V) - .0\overline{6}\,(P) - \dfrac{1}{20}\,(K) + .0\overline{3}\,(R) + .01\overline{6}\,(B + W) = T$

90. $.15\,(H) + \dfrac{1}{10}\,(R + I) + .075\,(B) + \dfrac{1}{20}\,(S + W) - .025\,(K + E) = T$

$.15\,(V) + \dfrac{1}{10}\,(P) + .075\,(K) - \dfrac{1}{20}\,(R) - .025\,(B + W) = T$

91. $-.15\,(H) - \dfrac{1}{10}\,(R + I) - .075\,(B) - \dfrac{1}{20}\,(S + W) + .025\,(K + E) = T$

$-.15\,(V) - \dfrac{1}{10}\,(P) - .075\,(K) + \dfrac{1}{20}\,(R) + .025\,(B + W) = T$

92. $\dfrac{3}{10}\,(H) + .2\,(R + I) + \dfrac{3}{20}\,(B) + .1\,(S + W) - \dfrac{1}{20}\,(K + E) = T$

$\dfrac{3}{10}\,(V) + .2\,(P) + \dfrac{3}{20}\,(K) - .1\,(R) - \dfrac{1}{20}\,(B + W) = T$

93. $-\dfrac{3}{10}\,(H) - .2\,(R + I) - \dfrac{3}{20}\,(B) - .1\,(S + W) + \dfrac{1}{20}\,(K + E) = T$

$\quad\;\; -\dfrac{3}{10}\,(V) - .2\,(P) - \dfrac{3}{20}\,(K) + .1\,(R) + \dfrac{1}{20}\,(B + W) = T$

94. $.6\,(H) + \dfrac{2}{5}\,(R + I) + .3\,(B) + \dfrac{1}{5}\,(S + W) - .1\,(K + E) = T$

$\quad\;\; .6\,(V) + \dfrac{2}{5}\,(P) + .3\,(K) - \dfrac{1}{5}\,(R) - .1\,(B + W) = T$

95. $-.6\,(H) - \dfrac{2}{5}\,(R + I) - .3\,(B) - \dfrac{1}{5}\,(S + W) + .1\,(K + E) = T$

$\quad\;\; -.6\,(V) - \dfrac{2}{5}\,(P) - .3\,(K) + \dfrac{1}{5}\,(R) + .1\,(B + W) = T$

96. $.9\,(H) + \dfrac{3}{5}\,(R + I) + \dfrac{9}{20}\,(B) + .3\,(S + W) - \dfrac{3}{20}\,(K + E) = T$

$\quad\;\; .9\,(V) + \dfrac{3}{5}\,(P) + \dfrac{9}{20}\,(K) - .3\,(R) - \dfrac{3}{20}\,(B + W) = T$

97. $-.9\,(H) - \dfrac{3}{5}\,(R + I) - \dfrac{9}{20}\,(B) - .3\,(S + W) + \dfrac{3}{20}\,(K + E) = T$

$\quad\;\; -.9\,(V) - \dfrac{3}{5}\,(P) - \dfrac{9}{20}\,(K) + .3\,(R) + \dfrac{3}{20}\,(B + W) = T$

Fractions with Positive Exponents

98. $\left(\dfrac{1}{2}\right)^{0}(H) + \left(\dfrac{1}{2}\right)^{1}(R + I) + \left(\dfrac{1}{2}\right)^{2}(B) + \left(\dfrac{1}{2}\right)^{3}(S + W) - \left(\dfrac{1}{2}\right)^{4}(K + E) = T$

$\quad\;\; \left(\dfrac{1}{2}\right)^{0}(V) + \left(\dfrac{1}{2}\right)^{1}(P) + \left(\dfrac{1}{2}\right)^{2}(K) - \left(\dfrac{1}{2}\right)^{3}(R) - \left(\dfrac{1}{2}\right)^{4}(B + W) = T$

99. $-\left(\dfrac{1}{2}\right)^{0}(H) - \left(\dfrac{1}{2}\right)^{1}(R + I) - \left(\dfrac{1}{2}\right)^{2}(B) - \left(\dfrac{1}{2}\right)^{3}(S + W) + \left(\dfrac{1}{2}\right)^{4}(K + E) = T$

$\quad\;\; -\left(\dfrac{1}{2}\right)^{0}(V) - \left(\dfrac{1}{2}\right)^{1}(P) - \left(\dfrac{1}{2}\right)^{2}(K) + \left(\dfrac{1}{2}\right)^{3}(R) + \left(\dfrac{1}{2}\right)^{4}(B + W) = T$

100. $\left(\dfrac{1}{3}\right)^{0}(H) + \left(\dfrac{1}{3}\right)^{1}(R + I) + \left(\dfrac{1}{3}\right)^{2}(B) + \left(\dfrac{1}{3}\right)^{3}(S + W) - \left(\dfrac{1}{3}\right)^{4}(K + E) = T$

$\quad\;\; \left(\dfrac{1}{3}\right)^{0}(V) + \left(\dfrac{1}{3}\right)^{1}(P) + \left(\dfrac{1}{3}\right)^{2}(K) - \left(\dfrac{1}{3}\right)^{3}(R) - \left(\dfrac{1}{3}\right)^{4}(B + W) = T$

101. $-\left(\dfrac{1}{3}\right)^{0}(H) - \left(\dfrac{1}{3}\right)^{1}(R + I) - \left(\dfrac{1}{3}\right)^{2}(B) - \left(\dfrac{1}{3}\right)^{3}(S + W) + \left(\dfrac{1}{3}\right)^{4}(K + E) = T$

$\quad\;\; -\left(\dfrac{1}{3}\right)^{0}(V) - \left(\dfrac{1}{3}\right)^{1}(P) - \left(\dfrac{1}{3}\right)^{2}(K) + \left(\dfrac{1}{3}\right)^{3}(R) + \left(\dfrac{1}{3}\right)^{4}(B + W) = T$

102. $\left(\dfrac{1}{4}\right)^{0}(H) + \left(\dfrac{1}{4}\right)^{1}(R + I) + \left(\dfrac{1}{4}\right)^{2}(B) + \left(\dfrac{1}{4}\right)^{3}(S + W) - \left(\dfrac{1}{4}\right)^{4}(K + E) = T$

$\left(\dfrac{1}{4}\right)^{0}(V) + \left(\dfrac{1}{4}\right)^{1}(P) + \left(\dfrac{1}{4}\right)^{2}(K) - \left(\dfrac{1}{4}\right)^{3}(R) - \left(\dfrac{1}{4}\right)^{4}(B + W) = T$

103. $-\left(\dfrac{1}{4}\right)^{0}(H) - \left(\dfrac{1}{4}\right)^{1}(R + I) - \left(\dfrac{1}{4}\right)^{2}(B) - \left(\dfrac{1}{4}\right)^{3}(S + W) + \left(\dfrac{1}{4}\right)^{4}(K + E) = T$

$-\left(\dfrac{1}{4}\right)^{0}(V) - \left(\dfrac{1}{4}\right)^{1}(P) - \left(\dfrac{1}{4}\right)^{2}(K) + \left(\dfrac{1}{4}\right)^{3}(R) + \left(\dfrac{1}{4}\right)^{4}(B + W) = T$

104. $\left(\dfrac{5}{6}\right)^{0}(H) + \left(\dfrac{4}{5}\right)^{1}(R + I) + \left(\dfrac{3}{4}\right)^{2}(B) + \left(\dfrac{2}{7}\right)^{3}(S + W) - \left(\dfrac{2}{8}\right)^{4}(K + E) = T$

$\left(\dfrac{5}{6}\right)^{0}(V) + \left(\dfrac{4}{5}\right)^{1}(P) + \left(\dfrac{3}{4}\right)^{2}(K) - \left(\dfrac{2}{7}\right)^{3}(R) - \left(\dfrac{2}{8}\right)^{4}(B + W) = T$

105. $-\left(\dfrac{5}{6}\right)^{0}(H) - \left(\dfrac{4}{5}\right)^{1}(R + I) - \left(\dfrac{3}{4}\right)^{2}(B) - \left(\dfrac{2}{7}\right)^{3}(S + W) + \left(\dfrac{2}{8}\right)^{4}(K + E) = T$

$-\left(\dfrac{5}{6}\right)^{0}(V) - \left(\dfrac{4}{5}\right)^{1}(P) - \left(\dfrac{3}{4}\right)^{2}(K) + \left(\dfrac{2}{7}\right)^{3}(R) + \left(\dfrac{2}{8}\right)^{4}(B + W) = T$

Fractions with Negative Exponents

106. $\left(\dfrac{1}{2}\right)^{-4}(H) + \left(\dfrac{1}{2}\right)^{-3}(R + I) + \left(\dfrac{1}{2}\right)^{-2}(B) + \left(\dfrac{1}{2}\right)^{-1}(S + W) - \left(\dfrac{1}{2}\right)^{-1}(K + E) = T$

$\left(\dfrac{1}{2}\right)^{-4}(V) + \left(\dfrac{1}{2}\right)^{-3}(P) + \left(\dfrac{1}{2}\right)^{-2}(K) - \left(\dfrac{1}{2}\right)^{-1}(R) - \left(\dfrac{1}{2}\right)^{-1}(B + W) = T$

107. $-\left(\dfrac{1}{2}\right)^{-4}(H) - \left(\dfrac{1}{2}\right)^{-3}(R + I) - \left(\dfrac{1}{2}\right)^{-2}(B) - \left(\dfrac{1}{2}\right)^{-1}(S + W) + \left(\dfrac{1}{2}\right)^{-1}(K + E) = T$

$-\left(\dfrac{1}{2}\right)^{-4}(V) - \left(\dfrac{1}{2}\right)^{-3}(P) - \left(\dfrac{1}{2}\right)^{-2}(K) + \left(\dfrac{1}{2}\right)^{-1}(R) + \left(\dfrac{1}{2}\right)^{-1}(B + W) = T$

108. $\left(\dfrac{1}{3}\right)^{-4}(H) + \left(\dfrac{1}{3}\right)^{-3}(R + I) + \left(\dfrac{1}{3}\right)^{-2}(B) + \left(\dfrac{1}{3}\right)^{-1}(S + W) - \left(\dfrac{1}{3}\right)^{-1}(K + E) = T$

$\left(\dfrac{1}{3}\right)^{-4}(V) + \left(\dfrac{1}{3}\right)^{-3}(P) + \left(\dfrac{1}{3}\right)^{-2}(K) - \left(\dfrac{1}{3}\right)^{-1}(R) - \left(\dfrac{1}{3}\right)^{-1}(B + W) = T$

109. $-\left(\dfrac{1}{3}\right)^{-4}(H) - \left(\dfrac{1}{3}\right)^{-3}(R + I) - \left(\dfrac{1}{3}\right)^{-2}(B) - \left(\dfrac{1}{3}\right)^{-1}(S + W) + \left(\dfrac{1}{3}\right)^{-1}(K + E) = T$

$-\left(\dfrac{1}{3}\right)^{-4}(V) - \left(\dfrac{1}{3}\right)^{-3}(P) - \left(\dfrac{1}{3}\right)^{-2}(K) + \left(\dfrac{1}{3}\right)^{-1}(R) + \left(\dfrac{1}{3}\right)^{-1}(B + W) = T$

110. $\left(\dfrac{1}{4}\right)^{-4}(H) + \left(\dfrac{1}{4}\right)^{-3}(R + I) + \left(\dfrac{1}{4}\right)^{-2}(B) + \left(\dfrac{1}{4}\right)^{-1}(S + W) - \left(\dfrac{1}{4}\right)^{-1}(K + E) = T$

$\left(\dfrac{1}{4}\right)^{-4}(V) + \left(\dfrac{1}{4}\right)^{-3}(P) + \left(\dfrac{1}{4}\right)^{-2}(K) - \left(\dfrac{1}{4}\right)^{-1}(R) - \left(\dfrac{1}{4}\right)^{-1}(B + W) = T$

111. $-\left(\dfrac{1}{4}\right)^{-4}(H) - \left(\dfrac{1}{4}\right)^{-3}(R + I) - \left(\dfrac{1}{4}\right)^{-2}(B) - \left(\dfrac{1}{4}\right)^{-1}(S + W) + \left(\dfrac{1}{4}\right)^{-1}(K + E) = T$

$-\left(\dfrac{1}{4}\right)^{-4}(V) - \left(\dfrac{1}{4}\right)^{-3}(P) - \left(\dfrac{1}{4}\right)^{-2}(K) + \left(\dfrac{1}{4}\right)^{-1}(R) + \left(\dfrac{1}{4}\right)^{-1}(B + W) = T$

112. $\left(\dfrac{2}{8}\right)^{-4}(H) + \left(\dfrac{2}{7}\right)^{-3}(R + I) + \left(\dfrac{3}{4}\right)^{-2}(B) + \left(\dfrac{4}{5}\right)^{-1}(S + W) - \left(\dfrac{5}{6}\right)^{-1}(K + E) = T$

$\left(\dfrac{2}{8}\right)^{-4}(V) + \left(\dfrac{2}{7}\right)^{-3}(P) + \left(\dfrac{3}{4}\right)^{-2}(K) - \left(\dfrac{4}{5}\right)^{-1}(R) - \left(\dfrac{5}{6}\right)^{-1}(B + W) = T$

113. $-\left(\dfrac{2}{8}\right)^{-4}(H) - \left(\dfrac{2}{7}\right)^{-3}(R + I) - \left(\dfrac{3}{4}\right)^{-2}(B) - \left(\dfrac{4}{5}\right)^{-1}(S + W) + \left(\dfrac{5}{6}\right)^{-1}(K + E) = T$

$-\left(\dfrac{2}{8}\right)^{-4}(V) - \left(\dfrac{2}{7}\right)^{-3}(P) - \left(\dfrac{3}{4}\right)^{-2}(K) + \left(\dfrac{4}{5}\right)^{-1}(R) + \left(\dfrac{5}{6}\right)^{-1}(B + W) = T$

Decimals with Positive Exponents

114. $3\,(H) + .3^1\,(R + I) + .3^2\,(B) + .3^3\,(S + W) - .3^4\,(K + E) = T$
$3\,(V) + .3^1\,(P) + .3^2\,(K) - .3^3\,(R) - .3^4\,(B + W) = T$

115. $-3\,(H) - .3^1\,(R + I) - .3^2\,(B) - .3^3\,(S + W) + .3^4\,(K + E) = T$
$-3\,(V) - .3^1\,(P) - .3^2\,(K) + .3^3\,(R) + .3^4\,(B + W) = T$

116. $4\,(H) + .4^1\,(R + I) + .4^2\,(B) + .4^3\,(S + W) - .4^4\,(K + E) = T$
$4\,(V) + .4^1\,(P) + .4^2\,(K) - .4^3\,(R) - .4^4\,(B + W) = T$

117. $-4\,(H) - .4^1\,(R + I) - .4^2\,(B) - .4^3\,(S + W) + .4^4\,(K + E) = T$
$-4\,(V) - .4^1\,(P) - .4^2\,(K) + .4^3\,(R) + .4^4\,(B + W) = T$

118. $5\,(H) + .5^1\,(R + I) + .5^2\,(B) + .5^3\,(S + W) - .5^4\,(K + E) = T$
$5\,(V) + .5^1\,(P) + .5^2\,(K) - .5^3\,(R) - .5^4\,(B + W) = T$

119. $-5\,(H) - .5^1\,(R + I) - .5^2\,(B) - .5^3\,(S + W) + .5^4\,(K + E) = T$
$-5\,(V) - .5^1\,(P) - .5^2\,(K) + .5^3\,(R) + .5^4\,(B + W) = T$

120. $6\,(H) + .6^1\,(R + I) + .6^2\,(B) + .6^3\,(S + W) - .6^4\,(K + E) = T$
$6\,(V) + .6^1\,(P) + .6^2\,(K) - .6^3\,(R) - .6^4\,(B + W) = T$

121. $-6\,(H) - .6^1\,(R + I) - .6^2\,(B) - .6^3\,(S + W) + .6^4\,(K + E) = T$
$-6\,(V) - .6^1\,(P) - .6^2\,(K) + .6^3\,(R) + .6^4\,(B + W) = T$

Decimals with Negative Exponents

122. $.3^{-4}\,(H) + .3^{-3}\,(R + I) + .3^{-2}\,(B) + .3^{-1}\,(S + W) - .3^{-1}\,(K + E) = T$
$.3^{-4}\,(V) + .3^{-3}\,(P) + .3^{-2}\,(K) - .3^{-1}\,(R) - .3^{-1}\,(B + W) = T$

123. $-.3^{-4}\,(H) - .3^{-3}\,(R + I) - .3^{-2}\,(B) - .3^{-1}\,(S + W) + .3^{-1}\,(K + E) = T$
$-.3^{-4}\,(V) - .3^{-3}\,(P) - .3^{-2}\,(K) + .3^{-1}\,(R) + .3^{-1}\,(B + W) = T$

124. $.4^{-4}(H) + .4^{-3}\,(R + I) + .4^{-2}(B) + .4^{-1}(S + W) - .4^{-1}(K + E) = T$
$.4^{-4}\,(V) + .4^{-3}\,(P) + .4^{-2}\,(K) - .4^{-1}\,(R) - .4^{-1}\,(B + W) = T$

125. $-.4^{-4}(H) - .4^{-3} (R + I) - .4^{-2}(B) - .4^{-1}(S + W) + .4^{-1} (K + E) = T$
$-.4^{-4}(V) - .4^{-3} (P) - .4^{-2}(K) + .4^{-1} (R) + .4^{-1}(B + W) = T$

126. $.5^{-4}(H) + .5^{-3} (R + I) + .5^{-2}(B) + .5^{-1} (S + W) - .5^{-1}(K + E) = T$
$.5^{-4}(V) + .5^{-3} (P) + .5^{-2} (K) - .5^{-1} (R) - .5^{-1}(B + W) = T$

127. $-.5^{-4}(H) - .5^{-3} (R + I) - .5^{-2}(B) - .5^{-1} (S + W) + .5^{-1}(K + E) = T$
$-.5^{-4}(V) - .5^{-3} (P) - .5^{-2} (K) + .5^{-1} (R) + .5^{-1}(B + W) = T$

128. $.6^{-4}(H) + .6^{-3} (R + I) + .6^{-2} (B) + .6^{-1} (S + W) - .6^{-1}(K + E) = T$
$.6^{-4}(V) + .6^{-3}(P) + .6^{-2} (K) - .6^{-1} (R) - .6^{-1} (B + W) = T$

129. $-.6^{-4}(H) - .6^{-3} (R + I) - .6^{-2} (B) - .6^{-1} (S + W) + .6^{-1}(K + E) = T$
$-.6^{-4}(V) - .6^{-3}(P) - .6^{-2} (K) + .6^{-1} (R) + .6^{-1} (B + W) = T$

Integers with Positive Exponents

130. $2^4 (H) + 2^3 (R + I) + 2^2 (B) + 2^1 (S + W) - 2^0 (K + E) = T$
$2^4 (V) + 2^3 (P) + 2^2 (K) - 2^1 (R) - 2^0 (B + W) = T$

131. $-2^4 (H) - 2^3 (R + I) - 2^2 (B) - 2^1 (S + W) + 2^0 (K + E) = T$
$-2^4 (V) - 2^3 (P) - 2^2 (K) + 2^1 (R) + 2^0 (B + W) = T$

132. $3^4 (H) + 3^3 (R + I) + 3^2 (B) + 3^1 (S + W) - 3^0 (K + E) = W$
$3^4 (V) + 3^3 (P) + 3^2 (K) - 3^1 (R) - 3^0 (B + W) = T$

133. $-3^4 (H) - 3^3 (R + I) - 3^2 (B) - 3^1 (S + W) + 3^0 (K + E) = T$
$-3^4 (V) - 3^3 (P) - 3^2 (K) + 3^1 (R) + 3^0 (B + W) = T$

134. $4^4 (H) + 4^3 (R + I) + 4^2 (B) + 4^1 (S + W) - 4^0 (K + E) = T$
$4^4 (V) + 4^3 (P) + 4^2 (K) - 4^1 (R) - 4^0 (B + W) = T$

135. $-4^4 (H) - 4^3 (R + I) - 4^2 (B) - 4^1 (S + W) + 4^0 (K + E) = T$
$-4^4 (V) - 4^3 (P) - 4^2 (K) + 4^1 (R) + 4^0 (B + W) = T$

136. $5^4 (H) + 5^3 (R + I) + 5^2 (B) + 5^1 (S + W) - 5^0 (K + E) = T$
$5^4 (V) + 5^3 (P) + 5^2 (K) - 5^1 (R) - 5^0 (B + W) = T$

137. $-5^4 (H) - 5^3 (R + I) - 5^2 (B) - 5^1 (S + W) + 5^0 (K + E) = T$
$-5^4 (V) - 5^3 (P) - 5^2 (K) + 5^1 (R) + 5^0 (B + W) = T$

138. $6^4 (H) + 6^3 (R + I) + 6^2 (B) + 6^1 (S + W) - 6^0 (K + E) = T$
$6^4 (V) + 6^3 (P) + 6^2 (K) - 6^1 (R) - 6^0 (B + W) = T$

139. $-6^4 (H) - 6^3 (R + I) - 6^2 (B) - 6^1 (S + W) + 6^0 (K + E) = T$
$-6^4 (V) - 6^3 (P) - 6^2 (K) + 6^1 (R) + 6^0 (B + W) = T$

Integers with Negative Exponents

140. $2 (H) + 2^{-1} (R + I) + 2^{-2} (B) + 2^{-3} (S + W) - 2^{-4} (K + E) = T$
$2 (V) + 2^{-1} (P) + 2^{-2} (K) - 2^{-3} (R) - 2^{-4} (B + W) = T$

141. $-2 (H) - 2^{-1} (R + I) - 2^{-2} (B) - 2^{-3} (S + W) + 2^{-4} (K + E) = T$
$-2 (V) - 2^{-1} (P) - 2^{-2} (K) + 2^{-3} (R) + 2^{-4} (B + W) = T$

142. $3 (H) + 3^{-1} (R + I) + 3^{-2} (B) + 3^{-3} (S + W) - 3^{-4} (K + E) = T$
$3 (V) + 3^{-1} (P) + 3^{-2} (K) - 3^{-3} (R) - 3^{-4} (B + W) = T$

143. $-3 (H) - 3^{-1} (R + I) - 3^{-2} (B) - 3^{-3} (S + W) + 3^{-4} (K + E) = T$
$-3 (V) - 3^{-1} (P) - 3^{-2} (K) + 3^{-3} (R) + 3^{-4} (B + W) = T$

144. $4 (H) + 4^{-1} (R + I) + 4^{-2} (B) + 4^{-3} (S + W) - 4^{-4} (K + E) = T$
$4 (V) + 4^{-1} (P) + 4^{-2} (K) - 4^{-3} (R) - 4^{-4} (B + W) = T$

145. $-4 (H) - 4^{-1} (R + I) - 4^{-2} (B) - 4^{-3} (S + W) + 4^{-4} (K + E) = T$
$-4 (V) - 4^{-1} (P) - 4^{-2} (K) + 4^{-3} (R) + 4^{-4} (B + W) = T$

146. $5 (H) + 5^{-1} (R + I) + 5^{-2} (B) + 5^{-3} (S + W) - 5^{-4} (K + E) = T$
$5 (V) + 5^{-1} (P) + 5^{-2} (K) - 5^{-3} (R) - 5^{-4} (B + W) = T$

147. $-5 (H) - 5^{-1} (R + I) - 5^{-2} (B) - 5^{-3} (S + W) + 5^{-4} (K + E) = T$
$-5 (V) - 5^{-1} (P) - 5^{-2} (K) + 5^{-3} (R) + 5^{-4} (B + W) = T$

148. $6 (H) + 6^{-1} (R + I) + 6^{-2} (B) + 6^{-3} (S + W) - 6^{-4} (K + E) = T$
$6 (V) + 6^{-1} (P) + 6^{-2} (K) - 6^{-3} (R) - 6^{-4} (B + W) = T$

149. $-6 (H) - 6^{-1} (R + I) - 6^{-2} (B) - 6^{-3} (S + W) + 6^{-4} (K + E) = T$
$-6 (V) - 6^{-1} (P) - 6^{-2} (K) + 6^{-3} (R) + 6^{-4} (B + W) = T$

Roots

150. $\sqrt{121} (H) + \sqrt{100} (R + I) + \sqrt{81} (B) + \sqrt{64} (S + W) - \sqrt{49} (K + E) = T$
$\sqrt{121} (V) + \sqrt{100} (P) + \sqrt{81} (K) - \sqrt{64} (R) - \sqrt{49} (B + W) = T$

151. $-\sqrt{121} (H) - \sqrt{100} (R + I) - \sqrt{81} (B) - \sqrt{64} (S + W) + \sqrt{49} (K + E) = T$
$-\sqrt{121} (V) - \sqrt{100} (P) - \sqrt{81} (K) + \sqrt{64} (R) + \sqrt{49} (B + W) = T$

152. $\sqrt{144} (H) + \sqrt{64} (R + I) + \sqrt{49} (B) + \sqrt{36} (S + W) - \sqrt{1} (K + E) = T$
$\sqrt{144} (V) + \sqrt{64} (P) + \sqrt{49} (K) - \sqrt{36} (R) - \sqrt{1} (B + W) = T$

153. $-\sqrt{144} (H) - \sqrt{64} (R + I) - \sqrt{49} (B) - \sqrt{36} (S + W) + \sqrt{1} (K + E) = T$
$-\sqrt{144} (V) - \sqrt{64} (P) - \sqrt{49} (K) + \sqrt{36} (R) + \sqrt{1} (B + W) = T$

154. $\sqrt[3]{125} (H) + \sqrt[3]{64} (R + I) + \sqrt[3]{27} (B) + \sqrt[3]{8} (S + W) - \sqrt[3]{1} (K + E) = T$
$\sqrt[3]{125} (V) + \sqrt[3]{64} (P) + \sqrt[3]{27} (K) - \sqrt[3]{8} (R) - \sqrt[3]{1} (B + W) = T$

155. $-\sqrt[3]{125} (H) - \sqrt[3]{64} (R + I) - \sqrt[3]{27} (B) - \sqrt[3]{8} (S + W) + \sqrt[3]{1} (K + E) = T$
$-\sqrt[3]{125} (V) - \sqrt[3]{64} (P) - \sqrt[3]{27} (K) + \sqrt[3]{8} (R) + \sqrt[3]{1} (B + W) = T$

156. $\sqrt{25} (H) + \sqrt[3]{64} (R + I) + \sqrt[4]{81} (B) + \sqrt[5]{32} (S + W) - \sqrt[6]{1} (K + E) = T$
$\sqrt{25} (V) + \sqrt[3]{64} (P) + \sqrt[4]{81} (K) - \sqrt[5]{32} (R) - \sqrt[6]{1} (B + W) = T$

157. $-\sqrt{25} (H) - \sqrt[3]{64} (R + I) - \sqrt[4]{81} (B) - \sqrt[5]{32} (S + W) + \sqrt[6]{1} (K + E) = T$
$-\sqrt{25} (V) - \sqrt[3]{64} (P) - \sqrt[4]{81} (K) + \sqrt[5]{32} (R) + \sqrt[6]{1} (B + W) = T$

Factorials and Summations

158. $6! \ (H) + 5! \ (R + I) + 4! \ (B) + 3! \ (S + W) - 2! \ (K + E) = T$
      $6! \ (V) + 5! \ (P) + 4! \ (K) - 3! \ (R) - 2! \ (B + W) = T$

159. $-6! \ (H) - 5! \ (R + I) - 4! \ (B) - 3! \ (S + W) + 2! \ (K + E) = T$
      $-6! \ (V) - 5! \ (P) - 4! \ (K) + 3! \ (R) + 2! \ (B + W) = T$

160. $\left(\sum_{j=1}^{6} j \right) (H) + \left(\sum_{j=1}^{5} j \right) (R + I) + \left(\sum_{j=1}^{4} j \right) (B) + \left(\sum_{j=1}^{3} j \right) (S + W)$

$- \left(\sum_{j=1}^{2} j \right) (K + E) = T$

$\left(\sum_{j=1}^{6} j \right) (V) + \left(\sum_{j=1}^{5} j \right) (P) + \left(\sum_{j=1}^{4} j \right) (K) - \left(\sum_{j=1}^{3} j \right) (R) - \left(\sum_{j=1}^{2} j \right) (B + W) = T$

161. $-\left(\sum_{j=1}^{6} j \right) (H) - \left(\sum_{j=1}^{5} j \right) (R + I) - \left(\sum_{j=1}^{4} j \right) (B) - \left(\sum_{j=1}^{3} j \right) (S + W)$

$+ \left(\sum_{j=1}^{2} j \right) (K + E) = T$

$-\left(\sum_{j=1}^{6} j \right) (V) - \left(\sum_{j=1}^{5} j \right) (P) - \left(\sum_{j=1}^{4} j \right) (K) + \left(\sum_{j=1}^{3} j \right) (R) + \left(\sum_{j=1}^{2} j \right) (B + W) = T$

Fractions, Decimals, Factorials, Summations, Exponents, Roots

162. $4! \ (H) + \sqrt[3]{64} \ (R + I) + \left(\dfrac{3}{4} \right)^{-2} (B) + .025 \ (S + W) - \left(\dfrac{1}{5} \right)^{3} (K + E) = T$

$4! \ (V) + \sqrt[3]{64} \ (P) + \left(\dfrac{3}{4} \right)^{-2} (K) - .025 \ (R) - \left(\dfrac{1}{5} \right)^{3} (B + W) = T$

163. $-4! \ (H) - \sqrt[3]{64} \ (R + I) - \left(\dfrac{3}{4} \right)^{-2} (B) - .025 \ (S + W) + \left(\dfrac{1}{5} \right)^{3} (K + E) = T$

$-4! \ (V) - \sqrt[3]{64} \ (P) - \left(\dfrac{3}{4} \right)^{-2} (K) + .025 \ (R) + \left(\dfrac{1}{5} \right)^{3} (B + W) = T$

164. $\left(\sum_{j=1}^{3} j \right) (H) + \left(\dfrac{2}{5} \right)^{-1} (R + I) + 2! \ (B) + \left(\dfrac{5}{6} \right)^{0} (S + W) - (\sqrt[4]{16})^{-2} (K + E) = T$

$\left(\sum_{j=1}^{3} j \right) (V) + \left(\dfrac{2}{5} \right)^{-1} (P) + 2! \ (K) - \left(\dfrac{5}{6} \right)^{0} (R) - (\sqrt[4]{16})^{-2} (B + W) = T$

165. $-\left(\sum_{j=1}^{3} j \right) (H) - \left(\dfrac{2}{5} \right)^{-1} (R + I) - 2! \ (B) - \left(\dfrac{5}{6} \right)^{0} (S + W) + (\sqrt[4]{16})^{-2} (K + E) = T$

$-\left(\sum_{j=1}^{3} j \right) (V) - \left(\dfrac{2}{5} \right)^{-1} (P) - 2! \ (K) + \left(\dfrac{5}{6} \right)^{0} (R) + (\sqrt[4]{16})^{-2} (B + W) = T$

Chapter Two

Explaining Fantasy Baseball and Mathematics to Students: Handouts

The following pages consist of student handouts that you can reproduce for your students. Table 2.1 provides a description of each handout. The handouts are also included in the *Fantasy Baseball and Mathematics Student Workbook,* which accompanies this teacher's guide. Thus, if your students are using the student workbook, you do not need to make copies of these handouts.

Table 2.1. Descriptions of Student Handouts

Handout Number	Description
1	Description and rules of the game
2	Roster sheet to help students keep track of the players on their team
3	Example of a box score
4	Step-by-step instructions on how to access statistics online
5	The default scoring system
6	Practice in computing points, using the default scoring system
7	Introduction of the default total points equation, which uses the numerical values of the default scoring system
8	Practice in computing points, using the default total points equation
9	Weekly scoring worksheet that uses the non-algebraic method to compute points
10	Weekly scoring worksheet that uses the algebraic method to compute points (that is, total points equations)
11	Score sheet on which students can record their weekly points for posting on the bulletin board
12	Stacked-bar graph that is used in conjunction with several practice worksheets and quizzes

Student handouts

Description and Rules

Fantasy Baseball and Mathematics is a game in which you create and manage a team of professional baseball players. Players earn points for hits, walks, stolen bases, home runs, runs scored, and runs batted in. Players lose points for striking out or making errors. Each week, you will find the sum of the points earned by your players. The object of the game is to accumulate the highest number of points.

How to Select Players

There are two options for selecting players. Your teacher will decide which option your class will use.

Option 1: Permanent Teams with Salary Cap. You have a salary cap of $40,000,000. This is the total amount you can spend on player values. Select 13 players for the positions listed in Table 1. Your instructor will provide you with a list of players and their values. Your teacher will also tell you whether you will select pitchers. If you do not select pitchers, the salary cap is $35,000,000. Table 1 lists the number of players selected at each position as well as the number of players in a starting lineup for each position.

Notice that you will select three substitute players (one infielder, one outfielder, and one pitcher) to use when a starting player gets injured or is performing poorly. You may select a player even if another student has chosen the same one.

Table 1.   Baseball Positions: Number to Be Selected and Number in a Starting Lineup

Position	Number to Be Selected	Number in Starting Lineup
First base (1B)	1	1
Second base (2B)	1	1
Third base (3B)	1	1
Shortstop (SS)	1	1
Catcher (C)	1	1
Outfield (OF)	4	3
Designated hitter (DH)	1	1
Pitcher (P)	2	1
Infield (IF)	1	

Description and Rules *(Cont'd.)*

Option 2: Different Teams Each Week. Each week, you will select one team. For example, you may decide to select your hometown team for the first week of the game. If you do, however, you will no longer be allowed to choose that team in later weeks because each team can be selected only once by each student during the course of the game. Unlike in option 1, you will compute points using team statistics rather than individual statistics. For example, if your team had a total of 16 hits, that is the number that would be used to compute points.

Trades

You may trade players if you selected players using option 1. For example, you may want to trade a second baseman for an outfielder. Consequently, you would insert your substitute infielder into the starting lineup to take the place of the second baseman you traded away. If you make a trade, it is important that you make the necessary changes to your fantasy team roster. Salary cap numbers do not apply to trades.

Injuries

If you cannot locate a player's name in the box scores, he is probably injured or taking a day off. If a player is declared out for the year and if you used option 1 to select players, you may use the portion of the salary cap you spent on that player to purchase another player. A list of injured players can be found in newspapers as well as online at www.fantasysportsmath.com or on other sports Web sites.

Student handouts

Fantasy Team Roster

Name of Fantasy Team _____   Team Owner _____

Position	Name	Team	Cost
First base			
Second base			
Third base			
Shortstop			
Outfield			
Outfield			
Outfield			
Outfield			
Designated hitter			
Pitcher			
Pitcher			
Catcher			
Infield			
Total			

How to Read Box Scores

Table 1. Bats-Rats Box Score

Bats	ab	r	h	rbi	bb	so	lob	avg
N Colt ss	5	0	0	0	0	1	2	.293
T Flyer cf	5	1	2	0	0	1	0	.260
F Vargas 1b	5	0	2	1	0	1	1	.385
J Macky rf	5	2	3	0	0	0	1	.287
A Cortez 3b	4	2	2	0	1	0	1	.298
T Joon dh	3	0	1	1	1	0	4	.318
C Flores lf	4	1	1	2	0	0	2	.257
G Hollis c	3	0	1	2	0	1	2	.161
a—M Vilipane ph-c	1	0	0	0	0	0	0	.293
M Roper 2b	3	0	0	0	1	0	1	.111
J Ezatz Jr pr-2b	0	0	0	0	0	0	0	.269
Totals	38	6	12	6	3	4	14	

BATTING: 2B—T Flyer (9, C Targe); J Macky (15, P Garcia). 3B—T Flyer (1, C Targe).
BASERUNNING: **SB**—C Flores (3, 2nd base off C Targe/J Blanco).
FIELDING: **E**—M Roper (1, ground ball).

Student handouts

How to Read Box Scores *(Cont'd.)*

Table 1. Bats-Rats Box Score *(Cont'd.)*

Rats	ab	r	h	rbi	bb	so	lob	avg
L Carter ss	4	1	1	0	1	1	4	.299
R Renault 2b	5	1	1	0	0	0	3	.275
R Brady rf	3	3	2	0	2	1	2	.300
J Martinez 3b	4	2	3	2	1	1	1	.328
S Cooke dh	4	1	3	5	1	0	1	.286
R Davis pr-dh	0	0	0	0	0	0	0	.308
J Blanco c	4	1	2	1	1	0	2	.297
J Smith 1b	3	0	0	0	0	0	4	.242
T Allen 1b	1	0	0	0	1	0	1	.232
B Johnson cf	4	0	2	1	1	1	4	.238
T Blake lf	4	0	1	0	0	0	2	.244
Totals	36	9	15	9	8	4	24	

BATTING: 2B—J Martinez (13, J Carrillo); R Brady (17, J Carrillo); S Cooke (19, L Deringer). **HR**—S Cooke (7, 7th inning off W Brown 1 on, 1 out).

Note: ab = at bats; r = **runs;** h = **hits;** rbi = **runs batted in;** bb = **bases on balls (i.e., walk);** so = **strikeouts;** lob = left on base; avg = average. Batting: 2B = double; 3B = triple; **HR** = **home run;** S = sacrifice. Baserunning: SB = stolen base. Items in bold will be used in the Fantasy Baseball and Math game. Fielding: **E** = **error.**

Let's look at the statistics for Sammy Cooke, the designated hitter for the Rats. In the box score in Table 1, next to his name, notice that he scored one run and had three hits, five runs batted in, one base on balls, and zero strikeouts. At the bottom of the box score, notice that he also hit a home run.

Now let's look at statistics for pitchers. In the pitchers' box score in Table 2, notice that Juan Carrillo pitched $6\frac{1}{3}$ innings. Carrillo gave up nine hits, six runs, six bases on balls, and he had three strikeouts.

Additional statistics you will keep track of include errors, stolen bases, and wins by a pitcher. When a pitcher wins a game, you will see a "W" next to his name in the box score. Notice that the winning pitcher for this game was T. Estrada of the Rats.

How to Read Box Scores *(Cont'd.)*

Table 2. Bats-Rats Box Score: Pitchers' Statistics

Bats	ip	h	r	er	bb	so	hr	era
J Carrillo	$6\frac{1}{3}$	9	6	6	6	3	0	3.56
W Brown (L, 2–1;)	0	2	1	1	1	0	1	2.57
M Reddmon	$\frac{2}{3}$	3	2	2	1	1	0	3.73
L Deringer	1	1	0	0	0	0	0	4.82

Rats	ip	h	r	er	bb	so	hr	era
C Targe	$5\frac{1}{3}$	10	6	6	1	3	0	4.53
P Garcia	$1\frac{1}{3}$	1	0	0	1	0	0	6.75
T Estrada (W, 1–1)	$\frac{2}{3}$	0	0	0	0	0	0	6.57
T Lloyd (H, 13)	$\frac{2}{3}$	0	0	0	1	1	0	2.64
S Wolly (S, 15)	1	1	0	0	0	0	0	1.05

Note: **ip** = **innings pitched**; **h** = **hits**; **r** = **runs**; er = earned runs; **bb** = **bases on balls (i.e., walk)**; **so** = **strikeouts**; hr = home runs; era = earned run average. L = loss; W = win; H = number of holds (i.e., the number of times a pitcher has entered a game in a save situation and left the game with his team leading); S = save.

Players who have stolen bases or committed errors will be listed in the "Baserunning" or "Fielding" sections. For example, notice that Carlos Flores stole a base for the Bats (his third of the year; if he had stolen two bases, it would have been listed like this: C. Flores 2 (3,4, 2nd base off C Targe/J Blanco; 3rd base off C Targe/J Blanco). Also notice that Matt Roper committed one error for the Bats.

Student handouts

How to Collect Data

Each week, you will use newspapers or online resources to collect data from one game in which each of the players in your starting lineup participated. You can choose the game that produced the best statistics for each player. Accessing data online is the quickest and easiest method. Statistics are also archived online so that you can still collect data if you have missed a week or two. To locate statistics online at www.fantasysportsmath.com, use the following steps:

a. Click the "Get Baseball Stats" link.

b. On the following page, use the calendar to select any day from the previous week.

c. Find a team one of your players participated in and click on the box score for that game. You can find the game during the previous week in which each of your players produced the best statistics.

How to Compute Points

Table 1.   Default Scoring System for Nonpitchers

The default scoring system that is listed below is normally used each week to determine the ranking of student teams in the game. Your teacher may choose a different scoring system that is more appropriate for you.

For Each:	Players Earn:		
Home run (HR)	$\frac{1}{2}$	or	.500
Run scored (R)	$\frac{1}{3}$	or	.333
Run batted in (RBI)	$\frac{1}{3}$	or	.333
Hit (H)	$\frac{1}{6}$	or	.167
Stolen base (SB)	$\frac{1}{7}$	or	.143
Base on balls (BB)	$\frac{1}{7}$	or	.143
Strikeout (SO)	$-\frac{1}{21}$	or	$-.048$
Error (E)	$-\frac{1}{21}$	or	$-.048$

Student handouts

How to Compute Points *(Cont'd.)*

Table 2. Default Scoring System for Pitchers

For Each:	Players Earn:		
Win (W)	$\frac{1}{2}$	or	.500
Inning pitched (IP)*	$\frac{1}{3}$	or	.333
Strikeout (SO)	$\frac{1}{6}$	or	.167
Run allowed (R)	$-\frac{1}{7}$	or	−.143
Hit allowed (H)	$-\frac{1}{21}$	or	−.048
Base on balls allowed (BB)	$-\frac{1}{21}$	or	−.048

Note: All decimals are rounded to the nearest thousandth.

*Rounded down to the nearest whole number—for example, $6\frac{2}{3}$ would be rounded down to 6.

Practice in Computing Points, Using the Default Scoring System

Use the following chart to compute points for the players listed in the Bats-Rats box score.

Player	Number of Home Runs $\times \frac{1}{2}$	Number of Runs Scored and RBIs $\times \frac{1}{3}$	Number of Hits $\times \frac{1}{6}$	Number of Stolen Bases and Bases on Balls $\times \frac{1}{7}$	Number of Strikeouts and Errors $\times \left(-\frac{1}{21}\right)$	Total Individual Points
Vargas						
Roper						
Martinez						
Carter						
Blanco						
Flores						
Flyer						
Brady						
Cooke						

Pitcher	Number of Wins $\times \frac{1}{2}$	Number of Innings Pitched $\times \frac{1}{3}$	Number of Strikeouts $\times \frac{1}{6}$	Number of Runs Allowed $\times \left(-\frac{1}{7}\right)$	Number of Hits and Bases on Balls Allowed $\times \left(-\frac{1}{21}\right)$	Total Individual Points
Carrillo						
Total team points (add the numbers in the "Total Individual Points" column):						

Default Total Points Equations

If you are using the default scoring system, you can use the equations shown on this sheet (the *default total points equations*). These equations use the same numerical values as the default scoring system to assign points to the players.

Default Total Points Equation for Nonpitchers

$$\frac{1}{2}(H) + \frac{1}{3}(R + I) + \frac{1}{6}(B) + \frac{1}{7}(S + W) - \frac{1}{21}(K + E) = T$$

H = number of home runs
R = number of runs scored
I = number of runs batted in
B = number of hits
S = number of stolen bases
W = number of bases on balls (walks)
K = number of strikeouts
E = number of errors
T = total points earned for one week for one player

Default Total Points Equation for Pitchers

$$\frac{1}{2}(V) + \frac{1}{3}(P) + \frac{1}{6}(K) - \frac{1}{7}(R) - \frac{1}{21}(B + W) = T$$

V = number of wins
P = number of innings pitched, rounded down to the nearest whole number
K = number of strikeouts
R = number of runs allowed
B = number of hits allowed
W = number of bases on balls allowed
T = total points earned for one week for one pitcher

Practice in Computing Points, Using the Default Total Points Equations

$$\frac{1}{2}(H) + \frac{1}{3}(R + I) + \frac{1}{6}(B) + \frac{1}{7}(S + W) - \frac{1}{21}(K + E) = T$$

Compute points earned for the following players listed in the Bats-Rats box score:

Felipe Vargas

$$\frac{1}{2}(0) + \frac{1}{3}(0 + 1) + \frac{1}{6}(2) + \frac{1}{7}(0 + 0) - \frac{1}{21}(1 + 0) = \frac{13}{21}$$

Matt Roper

Julio Martinez

Louie Carter

Jose Blanco

Practice in Computing Points, Using the Default Total Points Equations *(Cont'd.)*

$$\frac{1}{2}(H) + \frac{1}{3}(R + I) + \frac{1}{6}(B) + \frac{1}{7}(S + W) - \frac{1}{21}(K + E) = T$$

Carlos Flores

Tim Flyer

Ray Brady

Sammy Cooke

Juan Carrillo (pitcher) $\qquad \frac{1}{2}(V) + \frac{1}{3}(P) + \frac{1}{6}(K) - \frac{1}{7}(R) - \frac{1}{21}(B + W) = T$

Total team points: _____

Peer Signature: _____

Weekly Scoring Worksheet (Week _____)

Fill in the numerical values at the top of each column. Then fill in the scores for each of your players.

Nonpitchers

Player	Number of Home Runs × _____	Number of Runs Scored and RBI's × _____	Number of Hits × _____	Number of Stolen Bases and Bases on Balls × _____	Number of Strikeouts and Errors × _____	Total Individual Points
1B						
2B						
3B						
SS						
C						
OF						
OF						
OF						
DH						

Pitcher

	Number of Wins × _____	Number of Innings Pitched × _____	Number of Strikeouts × _____	Number of Runs Allowed × _____	Number of Hits and Bases on Balls Allowed × _____	Total Individual Points
P						
Total team points:						

Peer Signature: _____

Weekly Scoring Worksheet Using Total Points Equations (Week _____)

Write the total points equation you are using in the box below. Next, compute the points for each of your players, using the chart.

Player	Computation	Points
1B		
2B		
3B		
SS		
OF		
OF		
OF		
C		
DH		
P (Use equation for pitchers)		
Total team points:		

Total Points Week-by-Week

Team Name _____   Student Name _____

Player	Week 1	Week 2	Week 3	Week 4	Week 5	Week 6
Weekly Total						
Cumulative Total						

Total Points Week-by-Week *(Cont'd.)*

Team Name _____   Student Name _____

Player	Week 7	Week 8	Week 9	Week 10	Week 11	Week 12
Weekly Total						
Cumulative Total						

Total Points Week-by-Week *(Cont'd.)*

Team Name _____ Student Name _____

Player	Week 13	Week 14	Week 15	Week 16	Week 17	Week 18
Weekly Total						
Cumulative Total						

Student handouts

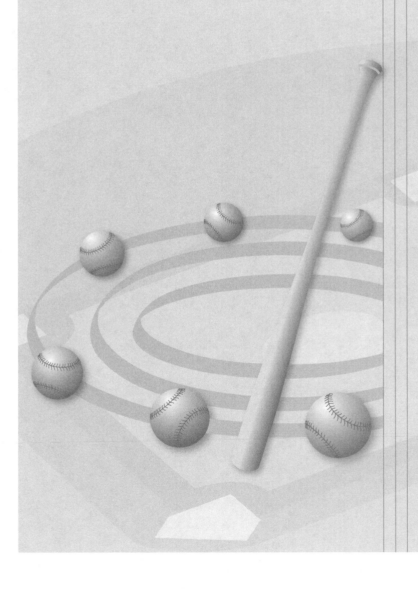

Using Graphs

Part Two

Chapter Three

Graphing Activities

Each week, students can construct circle, stacked-bar, or multiple-line graphs that reflect the performance of their teams. Students may create one or several graphs each week, depending on their skills. Although the graphs on the following pages are computer-generated, I highly recommend that students construct their graphs by hand so they can perform the computations themselves instead of relying on computer programs to perform the computations for them.

Circle Graphs

Circle graphs indicate the percentage of the fantasy team's points earned by each baseball player. The equation for computing the measurement of the central angle of a player's portion of the circle is as follows:

$$W \div S \times 360 = A$$

W = total weekly points for one player
S = total weekly points for the team
A = the measurement of the central angle of the circle graph

Figure 3.1. Circle Graph

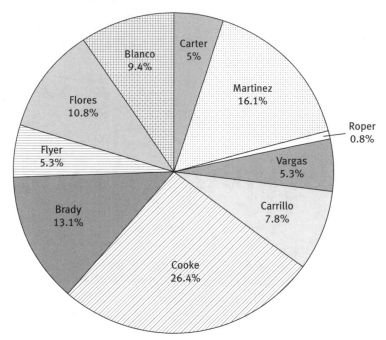

English Bulldogs Scoring Breakdown, Week 1

Example:

Julio Martinez's total points for week 1 (in simplest form): $1\frac{13}{14}$

Total points for the English Bulldogs for week 1: $11\frac{20}{21}$

Step 1: $1\frac{13}{14} \div 11\frac{20}{21} = \frac{27}{14} \div \frac{251}{21}$

Step 2: $\frac{27}{14} \div \frac{251}{21} = \frac{27}{14} \times \frac{21}{251}$

Step 3: $\frac{27}{14} \times \frac{21}{251} = \frac{81}{502} = .161$ (rounded to nearest thousandth)

Step 4: $.161 \times 360 = 57.96°$, which rounds to 58°

Figure 3.1 shows a circle graph of the points breakdown for the English Bulldogs based on their statistics from the box score in Tables 1.2 and 1.3.

Graphing activities

Stacked-Bar and Multiple-Line Graphs

A stacked-bar graph is a bar graph in which players' weekly points are "stacked" on top of each other. Multiple-line graphs are line graphs that depict the weekly points earned by two or more players. Examples of these graphs are found on the following pages. Points earned by individual players can be shown on stacked-bar graphs and multiple-line graphs. For example, students might choose three players and record those players' weekly points on stacked-bar and multiple-line graphs. Intervals of $\frac{2}{42}$ and $\frac{4}{42}$ work well for the stacked-bar and multiple-line graphs, assuming that students are using the default scoring system. Students may need to tape additional sheets of graph paper to the top of their first sheet to accommodate weeks in which their team scores significant points. My students constructed their graphs by hand, and I gave them extra credit if they created computer-generated charts.

The following pages contain examples of computer-generated graphs. Note that the stacked-bar graph is also used as a handout; its data is used as the basis for activities on several Practice Worksheets.

Stacked-Bar Graph

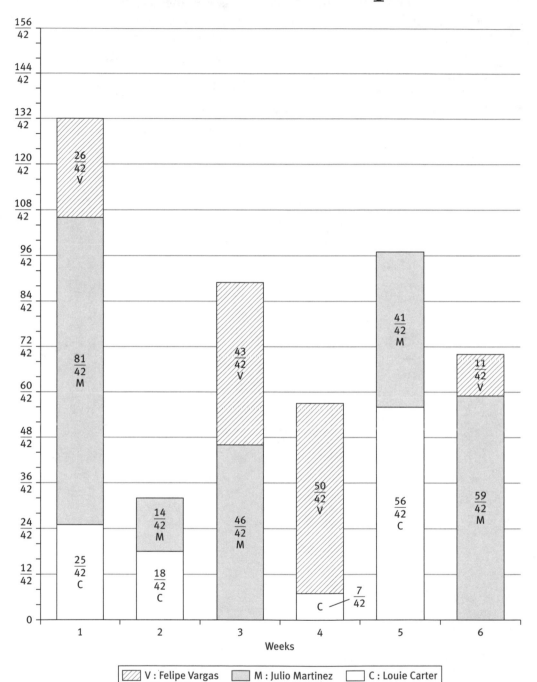

V : Felipe Vargas   M : Julio Martinez   C : Louie Carter

English Bulldogs Scoring Breakdown, Weeks 1–6

Graphing activities

Multiple-Line Graph

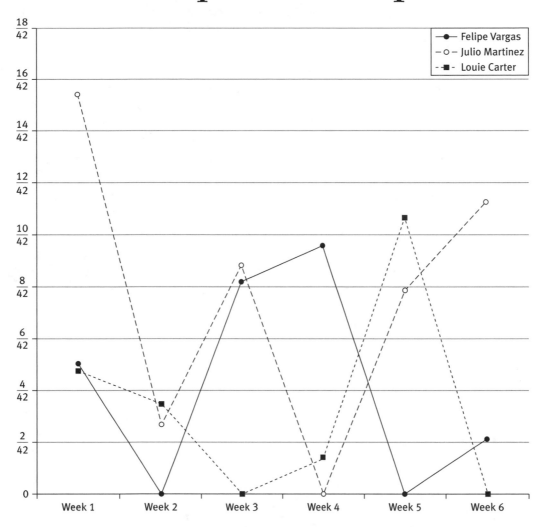

English Bulldogs Scoring Breakdown, Weeks 1–6

Graphing activities

57

Additional Options for Graphing Activities

- Graph all players' total points for one week on a non-stacked-bar graph in which each player is represented by his own column. Students can construct a new bar graph each week.

- Break down a stacked-bar graph according to positive and negative points earned. For each week, there would be two columns for each player: one column for the positive points (points earned) and another column for the negative points (points lost). The two columns could be next to each other, or the positive column could be above the x-axis and the negative column below the x-axis. (The latter scenario is also an option for multiple-line graphs).

- Select one or more scoring categories to graph. (This also works well for multiple-line graphs.) For instance, students could choose to graph hits, runs, or any other scoring criteria. Students would need to select appropriate intervals for the y-axis. Giving students opportunities to select appropriate intervals allows them to use higher-level thinking skills.

Part Three

Using Practice Worksheets and Quizzes

Chapter Four

Practice Worksheets

Practice worksheets from this book can be integrated into your existing curriculum in order to maximize the thematic approach to mathematics that the game provides. For example, if students are learning how to round decimals, you can use the corresponding worksheet in this text, in which students round the decimal equivalents for the points earned by their players. Consequently, students will be presented with opportunities to reinforce math concepts that tie in with the game. This approach will help students to make connections between math in school and math in the real world.

Each week, one or more worksheets may be integrated into your math lessons. Some worksheets (for example, Practice Worksheet 40: Mean, Median, Mode, Range) can be used for several weeks because students perform operations based on the points earned by their players for a given week. Other worksheets are used to compute cumulative points earned for the first few weeks of the season. These worksheets have a cumulative effect because students receive multiple exposures to the same material on a weekly basis. As the season progresses, students may work on several worksheets each week. Multiple exposures to content also facilitates mastery. For these reasons, it is highly recommended that students also participate in the graphing activities, for weekly exposures to circle, stacked-bar, and multiple-line graphs will help them to comprehend the material.

Students can also create their own worksheets based on their team's performance. For instance, if students are learning how to add and subtract fractions, they can write problems based on the points earned by their players—for example, "If player A earned three-eighths and player B earned four-fifths, how many total points did they earn?"

Rounding Whole Numbers and Expanded Notation

1. Round the following player salaries to the units given.

Salary	Nearest $10,000	Nearest $100,000	Nearest $1,000,000
$4,900,350	$4,900,000	$4,900,000	$5,000,000
$6,777,299			
$5,000,883			
$3,500,501			
$2,989,950			

2. Use expanded notation to represent the following player salaries.

Salary

$6,675,500 = $6,000,000 + $600,000 + $70,000 + $5,000 + $500

$6,444,700

$7,200,066

$4,950,222

$1,033,003

$5,999,999

Number sense

PRACTICE WORKSHEET 2

Least Common Multiple
and Greatest Common Factor

The number of home runs and runs batted in for five players during the first half of a season are listed below. Find the least common multiple and the greatest common factor for each pair of numbers.

	Number of Home Runs and RBIs	*Least Common Multiple*	*Greatest Common Factor*
Player A	22, 66	66	22
Player B	8, 36		
Player C	12, 42		
Player D	15, 40		
Player E	9, 30		

Number sense

PRACTICE WORKSHEET 3

Operations with Whole Numbers

1. What is the difference between the most expensive and least expensive players listed below?

Player A	$41,050,500
Player B	$16,555,935
Player C	$25,675,225
Player D	$38,000,810
Player E	$6,700,555

2. What is the total cost of the players listed in question 1?

3. What is the average cost of the players listed in question 1, to the nearest dollar?

4. If Charlie Mansfield had 39 hits in 13 games, how many hits did he average per game?

5. If 17 baseball players each have a salary of $3.5 million, what is the sum of their salaries?

Number sense

PRACTICE WORKSHEET 4

Equivalent Fractions

The points earned by players on the English Bulldogs are listed below. List the first three equivalent fractions for each.

Felipe Vargas $\dfrac{2}{3}$ $\dfrac{4}{6}$ $\dfrac{6}{9}$ $\dfrac{8}{12}$

Ray Brady $\dfrac{4}{7}$ _____ _____ _____

Julio Martinez $\dfrac{13}{42}$ _____ _____ _____

Louie Carter $\dfrac{3}{4}$ _____ _____ _____

Jose Blanco $\dfrac{5}{6}$ _____ _____ _____

Carlos Flores $\dfrac{5}{24}$ _____ _____ _____

Sammy Cooke $\dfrac{11}{21}$ _____ _____ _____

Number sense

PRACTICE WORKSHEET 5

Patterns and Multiples

(Use with Handout 12)

1. Find the first three multiples for the points earned by Julio Martinez for week 2. Reduce fractions to simplest form.

2. Find the first three multiples for the points earned by Louie Carter for week 4. Reduce fractions to simplest form.

3. Find the first three multiples for the points earned by Felipe Vargas for week 1. Reduce fractions to simplest form.

4. If $\frac{15}{21}$ is the third multiple of a number, what is the original number?

5. If 10.5 is the seventh multiple of a number, what is the original number?

Number sense

Name

PRACTICE WORKSHEET 6

Ordering Fractions and Decimals

(Use with Handout 12)

Example

For weeks 4–6, use inequalities to arrange the points earned by Felipe Vargas, Louie Carter, and Julio Martinez in ascending order:

$$\frac{7}{42} < \frac{11}{42} < \frac{41}{42} < 1\frac{8}{42} < 1\frac{14}{42} < 1\frac{17}{42}$$

After converting the fractions to decimals and rounding to the nearest thousandth, arrange the decimals in descending order:

$$1.405 > 1.333 > 1.190 > .976 > .262 > .167$$

1. Use inequalities and fractions to arrange in descending order the points earned by Felipe Vargas, Louie Carter, and Julio Martinez for weeks 1–3. Reduce fractions to simplest form.

2. Use inequalities and decimals to arrange in ascending order the points earned by Felipe Vargas, Louie Carter, and Julio Martinez for weeks 1–3. Round decimals to the nearest thousandth.

Number sense

67

PRACTICE WORKSHEET 7
Rounding Decimals

(Use with Handout 12)

Round players' cumulative points from weeks 1–6 to the nearest tenth, hundredth, and thousandth.

Example

A player's cumulative points for weeks 1–6 = $3\frac{4}{7}$ = 3.5714

Round to the nearest tenth = 3.6

Round to the nearest hundredth = 3.57

Round to the nearest thousandth = 3.571

Table 1

Player	Total Points in Weeks 1–6 (Fraction)	Total Points in Weeks 1–6 (Decimal)	Nearest Tenth	Nearest Hundredth	Nearest Thousandth
Carter					
Martinez					
Vargas					

In the following table, round the cumulative points for your three starting outfielders for weeks 1–6.

Table 2

Player	Total Points in Weeks 1–6 (Fraction)	Total Points in Weeks 1–6 (Decimal)	Nearest Tenth	Nearest Hundredth	Nearest Thousandth

Number sense

Improper Fractions, Mixed Numbers, and Reciprocals

The weekly point totals for a team are listed below. Convert all improper fractions to mixed numbers, and write them in their simplest form.

Example

$$\frac{72}{42} = 1\frac{30}{42} = 1\frac{5}{7}$$

1. $\dfrac{115}{42}$

2. $\dfrac{280}{42}$

3. $\dfrac{155}{42}$

4. $\dfrac{39}{42}$

5. $\dfrac{199}{42}$

Write the reciprocals (in simplest form) of the original fractions given in items 1–5.

6.

7.

8.

9.

10.

Number sense

69

PRACTICE WORKSHEET 9

Adding and Subtracting Fractions

(Use with Handout 12)

Example

For week 5, find the sum of the points earned by Felipe Vargas, Louie Carter, and Julio Martinez.

$$1\frac{14}{42} + \frac{41}{42} + 0 = 2\frac{13}{42}$$

1. For weeks 3 and 4, find the sum of the points earned by Felipe Vargas, Louie Carter, and Julio Martinez.

2. How many more points did the English Bulldogs earn during weeks 1–3 compared with weeks 4–6?

3. Find the sum of the points earned by Felipe Vargas and Julio Martinez for the first five weeks.

4. How many more cumulative points did Julio Martinez earn than Louie Carter during the six weeks?

5. How many fewer cumulative points did Felipe Vargas earn than Julio Martinez during the six weeks?

Number sense

Name

Stacked-Bar Graph

Using graph paper, use an interval of $\frac{4}{42}$ to construct a stacked-bar graph based on the data below. *Hint:* Convert all fractions so that they have a common denominator.

Player	Week 1	Week 2	Week 3
Carlos Hernandez	$\frac{1}{14}$	$\frac{3}{7}$	$\frac{29}{42}$
Bob Riley	$\frac{1}{3}$	$\frac{1}{6}$	$\frac{1}{42}$
Juan Alvarez	$\frac{5}{42}$	$\frac{8}{21}$	$\frac{1}{7}$

Multiplying and Dividing Fractions

(Use with Handout 12)

1. How many weeks would it take Johnny Allen to earn $5\frac{1}{7}$ points if he averaged $\frac{4}{7}$ points a week?

2. What is the product of the points earned by Louie Carter for weeks 4 and 5?

3. The product of the points earned by Steve Miller and Albert Trellast is $\frac{3}{7}$. If Miller earned $\frac{6}{7}$ points, how many points did Trellast earn?

4. Hal Young earned $17\frac{1}{4}$ points during the first 23 weeks. How many points did he average per week?

5. If Eric Hill earned 40.5 points for the season, and Ray Skidmore earned $\frac{1}{8}$ points per week, how many weeks would it take Skidmore to match Hill?

Number sense

Name _____

Rounding Fractions

(Use with Handout 12)

In Table 1, round players' cumulative points from weeks 1–6 to the nearest $\frac{1}{2}$, $\frac{1}{6}$, and $\frac{1}{7}$.

Example

Todd Walton's cumulative points for weeks 1–6 = $8\frac{10}{21}$

Round to the nearest $\frac{1}{2}$: $8\frac{1}{2}$

Round to the nearest $\frac{1}{6}$: $8\frac{3}{6}$

Round to the nearest $\frac{1}{7}$: $8\frac{3}{7}$

Table 1

	Nearest $\frac{1}{2}$	Nearest $\frac{1}{6}$	Nearest $\frac{1}{7}$
Vargas			
Carter			
Martinez			

Number sense

Rounding Fractions *(Cont'd.)*

In Table 2, round the cumulative points for your players for weeks 1–6.

Table 2

	Nearest $\frac{1}{2}$	Nearest $\frac{1}{6}$	Nearest $\frac{1}{7}$
1B			
2B			
3B			
SS			
C			
P			
DH			

Number sense

Name _____

Multiplying and Dividing Decimals

1. If an outfielder had annual player ratings of 65.89, 96.71, 76.75, 90.09, and 77.06, what would be his average rating for the last five years?

2. If a player worked 8 hours a day, 175 days a year, and his annual salary was $2.85 million, how much money did he make each working day? Each working hour? Each working minute? Each working second? Round your answers to the nearest cent.

3. If a snail can crawl at a rate of .04 yards per minute, how many hours will it take the snail to crawl from first base to second base (90 feet)? One mile?

4. If 27,000 fans each consumed an average of 10.55 ounces of soda at each game, how many ounces of soda were consumed for 5 games?

5. If a vendor selling ice cream sandwiches works 5 hours at $6.50 an hour and receives 33 cents for each sandwich sold, what is her income if she sold 297 sandwiches?

Number sense

Name _____

Unit Rates

Example

At a baseball game, you can purchase 16 oz. of soda for $2.95 or 24 oz. for $3.75. Which size is the lower price per ounce?

$$\$2.95 \div 16 \text{ oz.} = 18.4 \text{ cents per ounce}$$
$$\$3.75 \div 24 \text{ oz.} = 15.6 \text{ cents per ounce}$$

The 24-ounce size has the lower price per ounce.

1. You can purchase 12 oz. of peanuts for $2.50 or 20 oz. for $3.50. What is the lower price per ounce?

2. If Lance Richard drove his car 320 miles on 12 gallons of gas and Rich Hayes drove his car 380 miles on 15 gallons of gas, what is the mileage (miles per gallon) for each car?

3. If Luis Ayala can purchase 15 acres for $4.5 million or 20 acres for $6.5 million, what is the lower price per acre?

4. If you can purchase a season ticket (81 games) for $1,200 or a one-game ticket for $18, what is the lower price per game?

5. Which of the following provides the higher salary per year: $7.5 million for 8 years or $14.5 million for 15 years? What salary per year does it provide?

Number sense

Name

Converting Fractions, Decimals, and Percentages

(Use with Handout 12)

1. Find the cumulative points (weeks 1–6) for each player, and convert the fractions into decimals. Then round to the nearest tenth, hundredth, and thousandth. Finally, convert the decimal to a percentage, rounded to the nearest tenth.

Player	Total Points (Fraction)	Total Points (Decimal)	Rounded to Nearest Tenth	Rounded to Nearest Hundredth	Rounded to Nearest Thousandth	Percentage (Rounded to Nearest Tenth)
Vargas	$3\frac{2}{21}$	3.0952	3.1	3.10	3.095	309.5%
Carter						
Martinez						

2. Fill in the table below, using the cumulative points for the players on your team for the first six weeks.

Player	Total Points (Fraction)	Total Points (Decimal)	Rounded to Nearest Tenth	Rounded to Nearest Hundredth	Rounded to Nearest Thousandth	Percentage (Rounded to Nearest Tenth)
1B						
2B						
3B						
SS						
OF						
OF						
OF						
C						

Number sense

PRACTICE WORKSHEET 16

Ratios

(Use with Handout 12)

Example

Total points for Felipe Vargas to total points for Julio Martinez:

$$\frac{26}{42} \div \frac{81}{42} = \frac{26}{81} = .3209 = 32.1\%$$

For week 1, find the following ratios and convert them to percentages.

1. Total points for Julio Martinez to total points for Louie Carter

2. Total points for Louie Carter to total points for Felipe Vargas

Based on points earned in weeks 1–6, find the following ratio and convert to a percentage:

3. Total points for Felipe Vargas to total points for Julio Martinez

For week 6, find the following ratios and convert to percentages:

4. Total points for Julio Martinez to total points for Louie Carter and Felipe Vargas

5. Total points for Felipe Vargas to total points for Julio Martinez

Number sense

PRACTICE WORKSHEET 17

Percentage of Price Increase and Decrease

Example

If the price of a baseball jersey rose from $50 to $53, what is the percentage of price increase?

$$\frac{\text{Change in Price}}{\text{Original Price}} = \frac{3}{50} = .06 = 6\% \text{ increase}$$

1. If the price of an autographed baseball rose from $110 to $135, what is the percentage of price increase?

2. If the price of a baseball video game decreased from $55 to $49, what is the percentage of price decrease?

3. If the price of a season ticket decreased from $815 to $775, what is the percentage of price decrease?

4. If the price of a season ticket increased from $965 to $1,040, what is the percentage of price increase?

5. If the salary cap in Fantasy Baseball increased from $180 million to $200 million, what percentage increase would that represent?

Number sense

PRACTICE WORKSHEET 18

Finding a Percentage of a Number

Example

Rafael Cortes earned $1\frac{13}{14}$ points while Jose Gomez earned $\frac{13}{21}$ points. What percentage of Cortes's points do Gomez's points represent?

$$n \times 1\frac{13}{14} = \frac{13}{21}$$

$$\text{therefore, } n = \frac{13}{21} \div 1\frac{13}{14}$$

$$\text{thus, } n = \frac{26}{81} = .3210 = 32.1\%$$

1. Bobby Jackson earned $\frac{17}{42}$ points while Hal Gardner earned $3\frac{2}{21}$ points. What percentage of Gardner's points do Jackson's points represent?

2. Wally Price earned $1\frac{37}{42}$ points while Nick Franklin earned $\frac{5}{6}$ points. What percentage of Franklin's points do Price's points represent?

3. Nate Jones earned $5\frac{5}{21}$ points while Jeff Travis earned $8\frac{1}{42}$ points. What percentage of Jones's points do Travis's points represent?

4. Vern Yamamoto earned $2\frac{2}{3}$ points, which was 25% of Jimmy Smith's points. How many points did Smith earn?

Finding a Percentage of a Number *(Cont'd.)*

5. Grady O'Malley earned $\frac{5}{7}$ points, which was 71.428% of Adam Rice's points. How many points did Rice earn?

6. David Blakemore earned $2\frac{3}{4}$ points, which was 550% of Mariano Diaz's points. How many points did Diaz earn?

7. If Billy Dallimore earned 150% of $3\frac{1}{6}$, how many points did he earn?

8. There are 500,000 Bats fans in Chicago and 460,000 Rats fans in Boston. Each year, 10% of the Bats fans move to Boston, and 5% of the Rats fans move to Chicago. Complete the table below.

After year	Bats fans in Chicago	Rats fans in Boston
1		
2		
3		
4		

PRACTICE WORKSHEET 19

Proportions

Example

If Jeff Wood earned 2 points for his first 3 games of the season, how many points is he projected to earn for 30 games?

$$\frac{2}{3} = \frac{n}{30}$$

$$3(n) = 2(30)$$

$$3n = 60$$

$$n = 20$$

1. If Mark Salley earned $21\frac{5}{16}$ points during the first half of the season, how many points is he projected to earn for the entire season?

2. If it took Spencer Green six weeks to earn $4\frac{1}{2}$ points, how many weeks would it take him to earn $5\frac{1}{4}$ points?

3. If it took Greg Swarthmore 10 weeks to earn $16\frac{2}{3}$ points, how many weeks would it take him to earn 5 points?

4. Stan Fielding earned $26\frac{1}{4}$ points for 35 weeks. If he earned an equal amount of points each week, how many points would he have accumulated after eight weeks?

5. Barry Casterfield earned 45 points for 30 games. If he earned an equal amount of points for each game, how many points would he have accumulated after five games?

Number sense

Proportions *(Cont'd.)*

6. If Francisco Martinez had 56 strikeouts for the first seven games, how many strikeouts is he projected to get for 28 games?

7. Jake Ironsmith pitched 22 innings during his first three games. If he maintains his current pace, how many innings will he pitch in 31 games?

8. An architect is constructing a scale drawing of a new stadium. On the scale, one inch represents 40 feet. If the actual length from home plate to the right field foul post is 335 feet, what is the length from home plate to the foul post on the scale drawing?

9. In problem 8, what would be the actual height of the foul post if the scale drawing shows a height of 1.5 inches?

10. If it took 49 hours to drive 3,300 miles nonstop, how long would it take to drive 3,000 miles nonstop, assuming that the average speed would remain constant on both trips?

PRACTICE WORKSHEET 20

Ratios and Proportions

Example

The ratio of Zach Hiller's points to John Kenmore's points is 3:2. If Hiller earned 1.5 points, how many points did Kenmore earn?

$$\frac{3}{2} = \frac{1.5}{n}$$

$$3n = 2(1.5)$$

$$3n = 3$$

$$n = 1$$

1. The ratio of Ben Walker's points to Dante Johnson's points is 4:3. If Johnson earned .5 points, how many points did Walker earn?

2. The ratio of Curt Blood's points to Floyd Taylor's points is 1:3. If Blood earned $\frac{4}{7}$ points, how many points did Taylor earn?

3. The ratio of Mark Vultura's points to Max Luna's points is 5:2. If Luna earned 2.75 points, how many points did Vultura earn?

4. The ratio of Simon Nixon's points to Ken Sineway's points is 6:5. If Sineway earned 2.75 points, how many points did Nixon earn?

5. The ratio of Monte Anderson's points to Timmy McMillian's points is 13:21. If Anderson earned 4 points, how many points did McMillian earn?

Number sense

Factoring

Example

The product of the points earned for Sam Knight for weeks 1 and 2 is $3\frac{1}{6}$. If Knight earned $\frac{1}{3}$ points for week 1, how many points did he earn for week 2?

$$\frac{1}{3} \times n = \frac{19}{6}$$

$$n = \frac{19}{2} = 9\frac{1}{2}$$

1. The product of the points earned by Cliff Boozer and James Rillonia is $\frac{20}{42}$. If Boozer earned $\frac{5}{6}$ points, how many points did Rillonia earn?

2. Find two factors (other than 1) whose product equals $3\frac{1}{3}$.

3. The area of a rectangular parking lot is 200,000 square feet. If the length and width are whole numbers, what are the most realistic factors for the dimensions of the parking lot?

4. One factor of $\frac{39}{42}$ is $\frac{3}{7}$. Find a second factor.

5. If one factor of $\frac{3}{7}$ is 8, find a second factor.

Number sense

85

Interest, Depreciation, and Tax

1. If a player signed an eight-year contract for $120,000,000 and invested 25% of his annual salary at a rate of 5.25%, how much interest will he earn at the end of two years if the interest is compounded annually? (Assume that his income remains constant during the life of the contract.) Construct a spreadsheet showing the interest earned and the total value of his account at the end of each year. Use the following formula:

   $I = $ PRT
   $I = $ interest earned
   $P = $ principle
   $R = $ interest rate
   $T = $ time

2. If a player purchases a car for $190,000 and the state sales tax rate is 8.25%, how much tax will he owe? What will be the total cost of the car?

3. If the value of the automobile in problem 2 depreciates by 10% each year, what will the car be worth at the end of two years? Construct a spreadsheet showing the amount of depreciation and the corresponding value of the car each year.

4. If a player purchases a house for $4,250,000 and the price of the home appreciates 10% a year for the next two years, what will be the value of the home at the end of that period? Construct a spreadsheet showing the amount of annual appreciation and the corresponding value of the house at the end of each year.

Number sense

PRACTICE WORKSHEET 23

Prime Factorization

1. The weekly point totals (in 42nds) for a player are listed below. Write the prime factorization of each number, using exponents.

Week	Point Totals	Prime Factorization
Week 1	150	$2 \times 3 \times 5^2$
Week 2	99	
Week 3	7	
Week 4	64	
Week 5	125	
Week 6	0	
Week 7	19	

2. List the first five prime numbers: ____ ____ ____ ____ ____

PRACTICE WORKSHEET 24

Scientific Notation

The dimensions of a rectangular spring training facility are 1100 by 1500 feet. Write the area in scientific notation for the following units of measurement. *Hint:* 1 inch = 2.5 centimeters.

Example

Square feet: $1,100 \times 1,500 = 1,650,000$ sq. feet $= 1.65 \times 10^6$

1. Square inches:

2. Square yards:

3. Square centimeters:

4. Square millimeters:

5. Square meters:

Write the following in scientific notation.

6. 72.27

7. .000008

8. .900002

9. 142,887.511

10. 71,060,500.8

Write the following in standard form.

11. 33.005×10^{-2}

12. $16\frac{3}{4} \times 10^3$

Number sense

Name _____

Ordering Integers, Fractions, and Decimals

1. The following integers represent average December temperatures for several cities that host baseball teams. Place them in ascending order on the number line below.

    65   −4   −31   −14   −2   22   41   86   34   −1   61

    _____

2. The following integers represent the points earned for a player during the first ten games of the season. Place them in ascending order on the number line below.

    −38   22   37   71   −25   −13   −39   −51   91   −3

    _____

3. Place the following numerical values in ascending order on the number line below.

    $-.02258$ $8\frac{2}{5}$ $-4\frac{5}{16}$ -3.888 6.004 $-2\frac{7}{8}$

    _____

4. On the number line below, place the point totals earned by various players in ascending order.

    $\frac{3}{7}$ $\frac{13}{21}$ $\frac{1}{6}$ $.0235$ $\frac{5}{7}$ $.00235$ $.235$

    _____

Number sense

Name _____

Operations with Integers

1. If a player hit 504 home runs in 12 seasons, how many home runs did he average per season?

2. The numerical values below represent the points earned by 10 players on a team. How many points did the team gain or lose?

   −36   16   −38   29   110   −21   −49   −65   −23   57

3. If a player earned −26 points for the first half of the season, how many points is he projected to lose for a whole season?

4. If a player pitched 204 innings in 34 games, how many innings did he average per game?

Number sense

Operations with Integers *(Cont'd.)*

5. The numbers below represent profit or loss for five teams for one year. What is the average profit or loss?

    −$724,000 $9,987,500   −$2,722,311   −$116,766 $15,776,784

6. If one team lost $3,978,558 while another team reported a profit of $12,656,744, how much greater was the second team's profit than the other team's?

7. If one team reported a loss of $3,111,008, which included a profit of $2,777,456 on parking fees, how much money did it lose on operations other than the parking fees?

Name _____

Permutations and Combinations

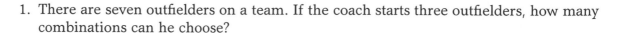

1. There are seven outfielders on a team. If the coach starts three outfielders, how many combinations can he choose?

2. If a team has jerseys in three different styles, caps in three different styles, and pants in two different styles, how many combinations of uniforms do they have to choose from?

3. A team's uniform consists of three colors, but they have five colors to choose from. How many combinations of uniforms do they have?

4. Before a game, five umpires line up in single file for the National Anthem. In how many ways can the umpires line up in single file?

Number sense

PRACTICE WORKSHEET 28

Unit Conversions

1. The distance between bases on a baseball diamond is 30 yards. What is this distance in inches? In centimeters? (2.5 centimeters = 1 inch)

2. If a player stole 45 bases in one year, how many feet did he run to steal those bases?

3. The distance from home plate to the left field foul post is 385 feet. What is this distance in yards? In inches?

4. The distance from home plate to the rubber on the pitcher's mound is 60′ 6″. What is this distance in centimeters? In millimeters? *Hint:* 10 mm = 1 cm

5. A team spent 2,765 minutes practicing last week. How many hours did they spend practicing last week?

6. A team is scheduled to play their next game in exactly 2 days, 14 hours. How many hours will pass before they play their next game? How many minutes?

Algebra and functions

Evaluating Algebraic Expressions

Evaluate $\frac{1}{2}(H) + \frac{1}{3}(R + I) + \frac{1}{6}(B) + \frac{1}{7}(S + W) - \frac{1}{21}(K + E)$ if

1. $H = 2$
    $R = 3$
    $I = 4$
    $B = 3$
    $S = 1$
    $W = 2$
    $K = 1$
    $E = 2$

2. $H = 1$
    $R = 2$
    $I = 5$
    $B = 4$
    $S = 0$
    $W = 1$
    $K = 2$
    $E = 1$

Evaluate $\frac{1}{2}(V) + \frac{1}{3}(P) + \frac{1}{6}(K) - \frac{1}{7}(R) - \frac{1}{21}(B + W)$ if

3. $V = 1$
    $P = 8$
    $K = 7$
    $R = 4$
    $B = 9$
    $W = 3$

Algebra and functions

Properties of Mathematics

Use numerical values to write one equation for each of the properties listed.

Distributive property $a(b + c) = ab + ac$

Commutative property of addition $a + b = b + a$

Commutative property of multiplication $ab = ba$

Associative property of addition $a + (b + c) = (a + b) + c$

Associative property of multiplication $a(bc) = (ab)c$

Inverse property of addition $a + (-a) = 0$

Inverse property of multiplication $a \times \dfrac{1}{a} = 1$

Identity property of addition $a + 0 = a$

Identity property of multiplication $a(1) = a$

Algebra and functions

Graphing on a Number Line

Example

During the first six weeks, Julio Martinez's range of points earned was between 0 and $1\frac{13}{14}$, inclusive. Using these data, we can graph the range of points earned by Martinez on a number line.

$$\overset{\bullet\rule[0.5ex]{6em}{0.4pt}\bullet}{\underset{0 \qquad\qquad 1\frac{13}{14}}{}}$$

Use a number line to graph the range of points earned by your players in weeks 1–6.

1B _____

2B _____

3B _____

SS _____

P _____

C _____

Linear Equations (A)

The equations below are used to compute total weekly points or to compute central angles in a circle graph. In each case, solve for the variable.

1. $\dfrac{1}{2}(H) + \dfrac{1}{3}(3 + 6) + \dfrac{1}{6}(3) + \dfrac{1}{7}(2 + 3) - \dfrac{1}{21}(1 + 1) = 5\dfrac{5}{42}$

2. $\dfrac{1}{6}(1) + \dfrac{1}{9}(2 + I) + \dfrac{1}{12}(4) + \dfrac{1}{18}(2 + 3) - \dfrac{1}{36}(2 + 1) = 1\dfrac{1}{4}$

3. $\dfrac{1}{9}(1) + \dfrac{1}{13.5}(R + 2) + \dfrac{1}{18}(2) + \dfrac{1}{27}(2 + 3) - \dfrac{1}{54}(3 + 1) = \dfrac{5}{9}$

4. $\dfrac{1}{10}(2) + \dfrac{1}{15}(3 + 7) + \dfrac{1}{20}(B) + \dfrac{1}{30}(0 + 1) - \dfrac{1}{60}(1 + 2) = 1\dfrac{1}{20}$

Algebra and functions

Linear Equations (A) *(Cont'd.)*

5. $w \div 10 \times 360 = 180$

6. $5^4(1) + 5^3(P) + 5^2(4) - 5^1(2) - 5^0(5 + 2) = 1{,}583$

7. $\dfrac{1}{4} \div s \times 360 = 30$

8. $.4^0(2) + .4^{-1}(2 + 1) + .4^{-2}(4) + .4^{-3}(0 + 1) - .4^{-4}(K + 2) = -106.125$

Algebra and functions

Linear Equations (A) *(Cont'd.)*

9. $\dfrac{w}{4.5} \times 360 = 90$

10. $.2\,(2) + .4\,(1 + 4) + .6(3) + .8\,(2 + W) - 1.2\,(3 + 1) = 2.6$

11. $.001\,(1) + .002\,(4 + 5) + .003\,(4) + .004\,(0 + 1) - .006\,(0 + E) = .017$

12. $\dfrac{3}{40}\,(0) + \dfrac{1}{20}\,(5 + 5) + .0375\,(4) + .025\,(S + 2) - .0125\,(1 + 1) = .7$

Algebra and functions

Linear Equations (A) *(Cont'd.)*

13. $\left(\dfrac{1}{3}\right)^0 (2) + \left(\dfrac{1}{3}\right)^1 (R + 1) + \left(\dfrac{1}{3}\right)^2 (4) + \left(\dfrac{1}{3}\right)^3 (3 + 0) - \left(\dfrac{1}{3}\right)^4 (2 + 1) = 3\dfrac{23}{27}$

14. $\left(\dfrac{1}{2}\right)^{-4} (V) + \left(\dfrac{1}{2}\right)^{-3} (5) + \left(\dfrac{1}{2}\right)^{-2} (6) - \left(\dfrac{1}{2}\right)^{-1} (4) - \left(\dfrac{1}{2}\right)^{0} (5 + 5) = 78$

15. $\sqrt[3]{125} (5) + \sqrt[3]{64} (8) + \sqrt[3]{27} (10) - \sqrt[3]{8} (1) - \sqrt[3]{1} (3 + W) = 60$

16. $\left(\displaystyle\sum_{j=1}^{6} j\right) (2) + \left(\displaystyle\sum_{j=1}^{5} j\right) (P) + \left(\displaystyle\sum_{j=1}^{4} j\right) (8) - \left(\displaystyle\sum_{j=1}^{3} j\right) (5) - \left(\displaystyle\sum_{j=1}^{2} j\right) (15 + 7) = 318$

17. $.2^{-3} (3) + \sqrt[3]{64} (0 + 2) + \left(\dfrac{3}{4}\right)^{-2} (0) + .025 (1 + 2) - \left(\dfrac{1}{4}\right)^{0} (K + 8) = 378.075$

Algebra and functions

Linear Equations (B)

In the problems below, insert the values shown for each variable in the total points equation. Then solve for T and write the answer in its simplest form.

$$\frac{1}{2}(H) + \frac{1}{3}(R + I) + \frac{1}{6}(B) + \frac{1}{7}(S + W) - \frac{1}{21}(K + E) = T$$

1. $H = 1$
   $R = 2$
   $I = 2$
   $B = 1$
   $S = 3$
   $W = 2$
   $K = 3$
   $E = 0$

2. $H = 0$
   $R = 4$
   $I = 2$
   $B = 3$
   $S = 0$
   $W = 1$
   $K = 1$
   $E = 2$

3. $H = 2$
   $R = 3$
   $I = 6$
   $B = 3$
   $S = 1$
   $W = 2$
   $K = 1$
   $E = 1$

Algebra and functions

Area and Perimeter of Rectangles

1. Explain the meaning of the variables in the following equations:

$$P = 2l + 2w$$

$$A = bh$$

2. The dimensions of a rectangular ballpark are 900 feet by 900 feet, and the dimensions of a rectangular store are 950 feet by 950 feet. Complete the table below. Do you see any patterns? Explain.

	Area of Ballpark	Area of Store	Ratio of Area of Ballpark to Area of Store
Square feet			
Square inches			
Square yards			
Square centimeters (2.5 cm = 1 inch)			
Square millimeters			
Square meters			

3. How much would it cost to resurface a baseball field if artificial turf costs $75 per square foot and the area that needs to be covered is 80,000 square feet?

Measurement and geometry

Area and Perimeter of Rectangles *(Cont'd.)*

4. Take a look at the player facilities below, then make two statements comparing their sizes. For example, you may predict that a dugout is 30 times smaller than a basketball court or that a soccer field is 400% larger than a basketball court. Then find the actual area and see how close your predictions were. Finally, find the perimeter of the player facilities.

Statement 1:

Statement 2:

Playing Area	Dimensions	Area	Perimeter
Dugout	60 ft. by 6 ft.		
Soccer field	55 m by 80 m		
Basketball court	85 ft. by 45 ft.		
Clubhouse	90 ft. by 38 ft.		

Name _____

Golden Rectangles

1. A Golden Rectangle is a rectangle in which the ratio of its length to its width is about 1.6:1. Fill in the chart below.

Playing Area	Dimensions	Ratio of Length to Width	Difference from Golden Rectangle Ratio
Clubhouse	90 ft. by 38 ft.		
Dugout	60 ft. by 6 ft.		
Basketball court	70 ft. by 42 ft.		
Soccer field	80 meters by 55 meters		

2. Which player areas have a ratio that approximates that of a Golden Rectangle?

3. Measure the length and width of various objects to find examples of Golden Rectangles. Some suggestions: flags, calculators, books, blackboards, windows, doors, file cabinets.

4. Predict the ratio of your height to the span of your two arms. Find the ratio. What did you learn?

Measurement and geometry

Name _____

Functions

In each of the following exercises, write the function rule and solve for the variable.

1. X = number of home runs; Y = points earned.

   Function rule: _____

X	Y
6	3
12	6
18	9
24	n

2. X = number of runs scored; Y = points earned.

   Function rule: _____

X	Y
5	$1\frac{2}{3}$
10	$3\frac{1}{3}$
15	5
16	n

Measurement and geometry

Functions *(Cont'd.)*

3. X = number of stolen bases; Y = number of points earned.

Function rule: _____

X	Y
7	1
14	2
21	3
35	n

4. Construct your own function chart below.

Function rule: _____

X	Y

Measurement and geometry

Name _____

Area and Circumference of Circles

Area of a circle $= \pi r^2$

Circumference of a circle $= \pi d$

r = radius; d = diameter; π = 3.14

1. A circular logo located at the center of a baseball field has a diameter of 29 feet. Find the area and circumference of the logo.

2. If the area of the on-deck circle is 28.26 square feet, what is the diameter of the circle?

3. The on-deck circle has a diameter of 3 feet. What is the area of the circle?

4. A circular baseball stadium has a radius of 800 feet. Find the diameter, circumference, and area of the stadium.

   Diameter: _____

   Circumference: _____

   Area: _____

5. If the circumference of a circular logo on a shirt is 3.5 inches, what are the radius, diameter, and area of the logo?

   Radius: _____

   Diameter: _____

   Area: _____

Measurement and geometry

Name _____

Weight

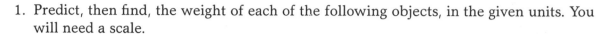

1. Predict, then find, the weight of each of the following objects, in the given units. You will need a scale.

	Predicted Weight			Actual Weight		
	Pounds	**Ounces**	**Grams**	**Pounds**	**Ounces**	**Grams**
Football						
Basketball						
Baseball						
Soccer ball						
Table tennis ball						
Hockey puck						

For each of the following problems, predict the answer, then solve the problem.

2. How many hockey pucks would weigh as much as five baseballs? As much as 500 table tennis balls?

3. Which is greater, the weight of four hockey pucks or six basketballs?

4. How many footballs would it take to equal your body weight? How many soccer balls? Baseballs?

Name _____

The Pythagorean Theorem

In a right triangle,

$$a^2 + b^2 = c^2$$

where

       a = length of one leg of the triangle
       b = length of the other leg of the triangle
       c = length of the hypotenuse

Use the Pythagorean Theorem to solve the following problems:

1. The distance between consecutive bases on a baseball diamond is 30 yards. Find the distance from home plate to second base.

2. Find the length of the diagonal of a parking lot if the length of the parking lot is 3,000 feet and the width is 1,800 feet.

3. Find the length of the diagonal of a clubhouse if the length of the clubhouse is 100 feet and the width is 60 feet.

4. Find the width of a dugout if the length of the dugout is 16 meters and the length of the diagonal is 17 meters.

5. Find the distance between the two foul poles at a stadium if both are 345 feet from home plate.

Measurement and geometry

PRACTICE WORKSHEET 40

Mean, Median, Mode, Range

(Use with Handout 12)

1. For weeks 4–6, find the points earned by each player on the English Bulldogs. In the table below, record the mean, median, mode, and range of those points earned for weeks 4–6.

Week	Mean	Median	Mode	Range
4				
5				
6				

2. For each of the first six weeks, find the individual score of each player on your starting team. In the table below, record the mean, median, mode, and range of those scores for each of the first six weeks.

Week	Mean	Median	Mode	Range
1				
2				
3				
4				
5				
6				

Name _____

PRACTICE WORKSHEET 41

Probability

1. Last year, Daniel Hopland sprayed 50% of his hits to the left side of the field, 20% to the right side, and 30% straightaway. If Hopland had 210 hits last year, how many hits did he have in each direction?

2. Using only the data in problem 1, find the probability that Hopland's first hit this year will be to the left side of the field.

3. A team's record during the last ten years is 900–720. Without taking any other variables into account, what should the team's record be this year?

4. Express numerically in several ways that the probability of an event occurring is 33%.

5. If the probability of an event occurring is 15%, what is the probability of that event not occurring?

Copyright © 2007 by Dan Flockhart

Statistics, data analysis, and probability   111

Probability *(Cont'd.)*

6. The letters in "zusu kiri choi" are placed into a hat. Find the probability of the following random events.

   A. Selecting the letter z

   B. Selecting the letters c, h, or s

   C. Selecting the letter i

   D. Selecting any letter except r

   E. Selecting the letter k, replacing it, then selecting the letter k again

   F. Selecting the letters o and u on consecutive draws (without replacing letters)

In exercises 7–11, you are given $P(Q)$, the probability that a player will hit a home run at various ballparks. Find $P(\text{not } Q)$, the probability that event Q will not occur.

7. $P(Q) = \dfrac{14}{42}$ $P(\text{not } Q) =$

8. $P(Q) = .395$ $P(\text{not } Q) =$

9. $P(Q) = 27\%$ $P(\text{not } Q) =$

10. $P(Q) = 1$ $P(\text{not } Q) =$

11. $P(Q) = 0$ $P(\text{not } Q) =$

Name _____

Circle Graphs

(Use with Handout 12)

$W \div S \times 360 = A$

W = total weekly points for one player
S = total weekly points for the team
A = the measurement of the central angle of the circle graph

Example

Julio Martinez earned $1\frac{13}{14}$ points for week 1. Find the measurement of the central angle in a circle graph representing Martinez's portion of his team's total points for week 1.

$$1\frac{13}{14} \div 3\frac{1}{7} \times 360 = 221°$$

1. Find the central angles for Felipe Vargas, Louie Carter, and Julio Martinez for week 5.

2. Find the central angles for Felipe Vargas, Louie Carter, and Julio Martinez for week 3.

3. Find the central angles showing the cumulative points for weeks 1–6 for each of the three English Bulldogs players.

4. If the central angle in a circle graph is 70 degrees, what percentage of the graph will that section represent?

5. If the central angle in a circle graph is 145 degrees, what percentage of the graph will that section represent?

6. If one section of a circle graph represents 62% of the total graph, what is the measurement of the corresponding central angle?

PRACTICE WORKSHEET 43

Stem-and-Leaf Plots and Histograms

1. The following values represent the point totals (in 42nds) for the English Bulldogs for the first 12 weeks of the season. Using graph paper, construct a stem-and-leaf plot and a histogram of the data.

   23   38   98   34   66   61

   52   88   85   77   73   49

2. The following values represent the point totals (in 42nds) for Sharp Listerman for the first 16 weeks of the season. Using graph paper, construct a stem-and-leaf plot and a histogram of the data.

   3   90   37   0   12   10   8   71

   17   66   0   29   4   0   16   37

Statistics, data analysis, and probability

PRACTICE WORKSHEET 44

Scatter Plots

1. The table below represents a player's age and number of home runs over the course of several seasons. On graph paper, construct a scatter plot of this data. Does the scatter plot have a positive or negative correlation? Explain.

Age	Number of Home Runs
30	57
31	55
32	49
33	45
34	46
35	44
36	39
37	32
38	26
39	19

2. The table below shows the number of hits and batting average for several players for one season. Using graph paper, construct a scatter plot of this data. Does the scatter plot have a positive or negative correlation? Explain.

Player	Number of Hits	Batting Average
A	87	.232
B	101	.240
C	133	.249
D	144	.251
E	145	.244
F	123	.267
G	232	.355
H	203	.328
I	187	.310

Statistics, data analysis, and probability

Box-and-Whisker Plots

The following data sets represent the points earned (in 42nds) by Jason Young and the total team points earned by the English Bulldogs for the last twelve weeks of the season. Using graph paper, draw a box-and-whisker plot for each set of data. On each plot, label the median as well as the upper and lower quartiles.

1. Jason Young   13, 0, 47, 3, 82, 10, 0, 21, 37, 4, 35, 39

2. The English Bulldogs   118, 81, 38, 50, 106, 77, 69, 111, 49, 43, 95, 133

Name _____

Statements Using Math Terminology

(Use with Handout 12)

Statements are complete sentences based on a set of data. Statements must be accompanied by mathematical proof. The statements on this worksheet were derived from Handout 12.

Example

Louie Carter earned half as many points in week 4 as Julio Martinez earned in week 2.

$$\frac{14}{42} \times \frac{1}{2} = \frac{14}{84} = \frac{7}{42}$$

For the following statements, show the mathematical proof.

1. Julio Martinez earned more cumulative points than Louie Carter and Felipe Vargas combined.

2. Louie Carter and Julio Martinez accounted for over 80% of the total points earned in week 1.

3. Felipe Vargas earned approximately 88% of the English Bulldogs' total points in week 4.

4. The difference between Felipe Vargas's and Louie Carter's cumulative points for weeks 4–6 was $\frac{1}{21}$.

5. Using the information from Handout 12, write five statements and show mathematical proof.

   A.

   B.

   C.

   D.

   E.

Mathematical reasoning 117

Name _____

Extra Credit Problems

1. Each time a basketball bounces, it rebounds to 72% of its height on the previous bounce. The ball was originally dropped from the top of a building that has a height of 150 feet. Construct a table of the number of bounces and the rebound height of each bounce. On which bounce does the ball bounce less than one foot high?

2. A. Find the stadium seating capacity, average ticket price, and revenue for two professional baseball teams. In this case, revenue is defined as the number of tickets sold (or seating capacity, if every game is sold out) multiplied by the average price of a ticket, multiplied by 81, which is the number of regular-season home games each team plays during one season.

   B. What is the difference in revenue between the two teams for one game? For one season?

   C. How much would revenue increase (for one game) for one team if they increased ticket prices by an average of 5%?

   D. How much would revenue decrease (for the season) for one team if they decreased ticket prices by an average of 3%?

3. Create a new scoring system, using fractions, decimals, exponents, factorials, roots, integers, or summations. Then compute your weekly points in that scoring system.

4. Predict how many small (8-inch diameter), medium (12-inch diameter), or large (16-inch diameter) pizzas would fit on a baseball field. Then find the actual amount.

Chapter Five

Quizzes

The quizzes in this chapter can be used to analyze student progress and provide timely feedback in order to improve student achievement. They also provide students with multiple exposures to content.

Research indicates that learning is maximized when assessment closely monitors the learning objectives. Consequently, the quizzes in this chapter are aligned with the corresponding worksheets. They can be used to assess student progress or they can be used as additional worksheets.

Research also indicates that learning is facilitated when visual materials are included in the learning environment. The inclusion of visual materials (in the form of charts, diagrams, or other data depiction) lessens the cognitive load on students. This is particularly true for special needs students. In order to take advantage of this research, several quizzes have been designed to be used in conjunction with Handout 12, a stacked-bar graph.

Some quizzes may be time-consuming, depending on the skill level of students (for example, the quizzes in which students construct graphs).

QUIZ 1

Rounding Whole Numbers and Expanded Notation

1. Round the following player salaries to the units given.

Salary	Nearest $10,000	Nearest $100,000	Nearest $1,000,000
$3,333,437			
$1,909,199			
$3,555,500			
$5,666,902			
$4,456,123			

2. Use expanded notation to represent the following player salaries.

Salary

$4,001,001

$6,449,501

$3,888,888

$3,100,704

$9,999,999

Number sense

QUIZ 2

Least Common Multiple
and Greatest Common Factor

Listed below are the numbers of home runs and runs batted in for five players during the first half of a season. Find the least common multiple and the greatest common factor for each pair of numbers.

	Number of Home Runs and RBIs	Least Common Multiple	Greatest Common Factor
Player A	35, 105		
Player B	4, 9		
Player C	29, 87		
Player D	6, 14		
Player E	7, 8		

Number sense

QUIZ 3

Operations with Whole Numbers

1. What is the difference between the most expensive and least expensive players listed below?

   Player A $42,700,885
   Player B $17,555,935
   Player C $25,675,225
   Player D $38,000,810
   Player E $5,700,250

2. What is the total cost of the players in problem 1?

3. What is the average cost of the players in problem 1, to the nearest dollar?

4. If a player had 450 hits in 300 games, how many hits did he average per game?

5. If 20 baseball players each have a salary of $2.75 million, what is the sum of their salaries?

Number sense

QUIZ 4

Equivalent Fractions

List the first three equivalent fractions for each player's points.

1. Felipe Vargas $\dfrac{3}{4}$ _____   _____   _____

2. Ray Brady $\dfrac{3}{7}$ _____   _____   _____

3. Julio Martinez $\dfrac{17}{21}$ _____   _____   _____

4. Louie Carter $\dfrac{8}{12}$ _____   _____   _____

5. Jose Blanco $\dfrac{8}{21}$ _____   _____   _____

6. Carlos Flores $\dfrac{3}{8}$ _____   _____   _____

7. Sammy Cooke $\dfrac{5}{6}$ _____   _____   _____

Number sense

QUIZ 5

Patterns and Multiples

(Use with Handout 12)

1. Find the first three multiples for the points earned by Julio Martinez for week 5. Reduce fractions to simplest form.

2. Find the first three multiples for the points earned by Louie Carter for week 2. Reduce fractions to simplest form.

3. Find the first three multiples for the points earned by Felipe Vargas for week 6. Reduce fractions to simplest form.

4. If $\frac{9}{7}$ is the third multiple of a number, what is the original number?

5. If 5 is the fourth multiple of a number, what is the original number?

QUIZ 6

Ordering Fractions
and Decimals

(Use with Handout 12)

1. For weeks 2–4, use inequalities and fractions to arrange the points earned by Felipe Vargas, Louie Carter, and Julio Martinez in ascending order. Reduce fractions to simplest form.

2. Convert the fractions from problem 1 to decimals, and round them to the nearest thousandth. Then arrange the decimals in descending order.

3. For weeks 1–5, use inequalities and fractions to arrange the points earned by Felipe Vargas and Louie Carter in ascending order. Reduce fractions to simplest form.

4. Convert the fractions from problem 3 to decimals, and round them to the nearest thousandth. Then arrange the decimals in descending order.

Number sense

QUIZ 7

Rounding Decimals

(Use with Handout 12)

1. Round players' cumulative points from weeks 2–5 to the nearest tenth, hundredth, and thousandth.

Player	Total Points in Weeks 2–5 (Fraction)	Total Points in Weeks 2–5 (Decimal)	Nearest Tenth	Nearest Hundredth	Nearest Thousandth
Carter					
Martinez					
Vargas					

2. In the following table, round the cumulative points for your three starting outfielders for weeks 2–5.

Player	Total Points in Weeks 2–5 (Fraction)	Total Points in Weeks 2–5 (Decimal)	Nearest Tenth	Nearest Hundredth	Nearest Thousandth

Number sense

QUIZ 8

Improper Fractions, Mixed Numbers, and Reciprocals

Convert the following points earned to mixed numbers, and write them in their simplest form.

1. $\dfrac{63}{42}$

2. $\dfrac{172}{42}$

3. $\dfrac{84}{42}$

4. $\dfrac{101}{42}$

5. $\dfrac{226}{42}$

Write the reciprocals (in simplest form) for the original fractions given above in items 1–5.

6.

7.

8.

9.

10.

Number sense

QUIZ 9

Adding and Subtracting Fractions

(Use with Handout 12)

1. For weeks 1–3, find the sum of the points earned by Felipe Vargas, Louie Carter, and Julio Martinez.

2. How many fewer points did the English Bulldogs earn during weeks 4–6 compared with weeks 1–3?

3. Find the sum of the points earned by Felipe Vargas and Julio Martinez for the first four weeks.

4. How many fewer cumulative points did Louie Carter earn than Julio Martinez during the six weeks?

5. How many more cumulative points did Julio Martinez earn than Felipe Vargas during the six weeks?

Number sense

QUIZ 10

Stacked-Bar Graph

Using graph paper, construct a stacked-bar graph based on the data below. *Hint:* Convert all fractions so that they have a common denominator.

Player	Week 1	Week 2	Week 3
A. Patrick Wilson	$\frac{2}{21}$	$\frac{1}{6}$	$\frac{8}{21}$
B. Joshua Morris	$\frac{3}{7}$	$\frac{5}{14}$	0
C. Gregory Ward	$\frac{5}{42}$	$\frac{1}{7}$	$\frac{3}{7}$

Number sense

QUIZ 11
Multiplying and Dividing Fractions

(Use with Handout 12)

1. How many weeks would it take Johnny Nelson to earn $1\frac{1}{3}$ points if he averaged $\frac{2}{21}$ points a week?

2. What is the product of the points earned by Louie Carter for weeks 2 and 5?

3. The product of the points earned by Barry Phillips and Pedro Sanchez is $\frac{17}{42}$. If Phillips earned $\frac{1}{2}$ points, how many points did Sanchez earn?

4. Michael White earned $22\frac{1}{2}$ points during the first 30 weeks. How many points did he average per week?

5. If Eric Martin earned $2\frac{4}{7}$ points in the first week while Arthur Torres earned $\frac{3}{14}$ points, how many weeks would it take Torres to match Martin's total for the first week?

Number sense

QUIZ 12
Rounding Fractions

(Use with Handout 12)

In the table below, round each player's cumulative points from weeks 2–5 to the nearest $\frac{1}{2}$, $\frac{1}{6}$, and $\frac{1}{7}$.

Table 1

	Nearest $\frac{1}{2}$	Nearest $\frac{1}{6}$	Nearest $\frac{1}{7}$
Vargas			
Carter			
Martinez			

In the following table, round the cumulative points for your players for weeks 2–5.

Table 2

	Nearest $\frac{1}{2}$	Nearest $\frac{1}{6}$	Nearest $\frac{1}{7}$
1B			
2B			
3B			
SS			
C			
P			
DH			

Number sense

QUIZ 13

Multiplying and Dividing Decimals

1. If an outfielder had annual player ratings of 77.04, 65.71, 76.75, 90.09, and 88.08, what would be his average rating for the last five years?

2. If a player worked 8 hours a day, 175 days a year, and his annual salary was $3.15 million, how much money did he make each working day? Each working hour? Each working minute? Each working second? Round your answers to the nearest cent.

3. If a snail can crawl at a rate of .02 yards per minute, how many hours will it take the snail to crawl from first base to second base (90 feet)? One mile?

4. If 29,700 fans each consumed an average of 9.25 ounces of soda at each game, how many ounces of soda were consumed for 10 games?

5. If a vendor selling ice cream sandwiches works 4 hours at $6.75 an hour and receives 32 cents for each sandwich sold, what is her income if she sold 305 sandwiches?

Number sense

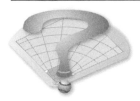

QUIZ 14

Unit Rates

1. You can purchase 12 oz. of peanuts for $2.85 or 18 oz. for $3.45. What is the lower price per ounce?

2. If Jerry Turner drove his car 405 miles on 14 gallons of gas and Walt Wright drove his car 365 miles on 11.5 gallons of gas, what is the mileage (miles per gallon) for each car?

3. If Larry Roberts can purchase 65 acres for $17.5 million or 45 acres for $6.25 million, what is the lower price per acre?

4. If you can purchase a season ticket (81 games) for $1,350 or a one-game ticket for $17, what is the lower price per game?

5. What is the higher salary per year of the following two: $9.25 million for six years or $10.5 million for seven years?

Converting Fractions, Decimals, and Percentages

(Use with Handout 12)

1. Find the cumulative points for weeks 1–3 for each player, and convert the fractions into decimals. Then round to the nearest tenth, hundredth, and thousandth. Finally, convert the decimal to a percentage, rounded to the nearest tenth.

Player	Total Points (Fraction)	Total Points (Decimal)	Rounded to Nearest Tenth	Rounded to Nearest Hundredth	Rounded to Nearest Thousandth	Percentage (Rounded to Nearest Tenth)
Vargas						
Carter						
Martinez						

2. Fill in the table below, using the cumulative points for the players on your team for weeks 1–3.

Player	Total Points (Fraction)	Total Points (Decimal)	Rounded to Nearest Tenth	Rounded to Nearest Hundredth	Rounded to Nearest Thousandth	Percentage (Rounded to Nearest Tenth)
1B						
2B						
3B						
SS						
OF						
OF						
OF						
C						

Number sense

QUIZ 16
Ratios

(Use with Handout 12)

Based on points earned in week 2, find the following ratios and convert them to percentages:

1. Total points for Julio Martinez to total points for Louie Carter

2. Total points for Louie Carter to total points for Felipe Vargas

Based on points earned in weeks 1–3, find the following ratio:

3. Total points for Felipe Vargas to total points for Julio Martinez

For week 5, find the following ratio:

4. Total points for Julio Martinez to total points for Louie Carter

For week 6, find the following ratio:

5. Total points for Felipe Vargas to total points for Julio Martinez

Number sense

QUIZ 17

Percentage of Price Increase and Decrease

1. If the price of an autographed baseball rose from $135 to $165, what is the percentage of price increase?

2. If the price of a baseball video game decreased from $35 to $32, what is the percentage of price decrease?

3. If the price of a season ticket decreased from $950 to $875, what is the percentage of price decrease?

4. If the price of a season ticket increased from $790 to $840, what is the percentage of price increase?

5. If the salary cap in Fantasy Baseball increased from $175 million to $190 million, what percentage increase would that represent?

Number sense

QUIZ 18

Finding a Percentage of a Number

1. Manuel Gomez earned $\frac{5}{6}$ points, while William Hall earned $1\frac{3}{7}$ points. What percentage of Hall's points do Gomez's points represent?

2. Carlos Lopez earned $3\frac{17}{42}$ points, while George Campbell earned $1\frac{19}{21}$ points. What percentage of Campbell's points do Lopez's points represent?

3. Harold Thomas earned $16\frac{2}{3}$ points, while Jeff Clark earned $7\frac{1}{7}$ points. What percentage of Thomas's points do Clark's points represent?

4. Juan Trejo earned $\frac{1}{2}$ points, which was 75% of Anthony Jones's points. How many points did Jones earn?

Number sense

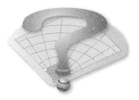

QUIZ 19

Proportions

1. Ronald Moore earned 15 points for 20 weeks. If he earned an equal amount of points each week, how many points would he have accumulated after five weeks?

2. If Paul Baker had 85 strikeouts for the first nine games, then how many strikeouts is he projected to get for 27 games?

3. Jimmy Anderson pitched 28 innings during his first four games. If he maintains his current pace, how many innings will he pitch in 20 games?

4. If Mark Harris earned $19\frac{17}{21}$ points for the first half of the season, how many points is he projected to earn for the entire season?

5. If it took Jason Rodriguez four weeks to earn $1\frac{5}{7}$ points, how many weeks would it take him to earn $4\frac{2}{7}$ points?

6. If it took Gary Evans five weeks to earn .3 points, how many weeks would it take him to earn $1\frac{1}{2}$ points?

Number sense

QUIZ 20

Ratios and Proportions

1. The ratio of Ben Walker's points to Danny Bolt's points is 5:2. If Bolt earned $\frac{1}{3}$ point, how many points did Walker earn?

2. The ratio of John Robinson's points to Curt Peterson's points is 2:3. If Robinson earned .5 point, how many points did Peterson earn?

3. The ratio of Tim Myers's points to Andy West's points is 2:7. If West earned 5 points, how many points did Myers earn?

4. The ratio of Troy Willis's points to Keith Diaz's points is 4:3. If Diaz earned 4 points, how many points did Willis earn?

5. The ratio of Jack Harrison's points to Matt Roper's points is 7:8. If Harrison earned 3.5 points, how many points did Roper earn?

Number sense

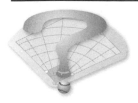

QUIZ 21

Factoring

1. The product of the points earned by two players is 1. If one player earned $\frac{7}{6}$ points, how many points did the second player earn?

2. Find two factors (other than 1) whose product equals $3\frac{3}{7}$.

3. The area of a rectangular parking lot is 180,000 square feet. If the length and width are whole numbers, what are some realistic factors for the dimensions of the parking lot?

4. One factor of $\frac{16}{21}$ is $\frac{2}{7}$. Find a second factor.

5. If one factor of $\frac{36}{45}$ is 6, find a second factor.

Number sense

QUIZ 22

Interest, Depreciation, and Tax

1. If a player signs a five-year contract for $75,000,000 and invests 15% of his annual salary at a rate of 4.25%, how much interest will he earn at the end of two years if the interest is compounded annually? (Assume that his income remains constant during the life of the contract.) Construct a spreadsheet showing the interest earned and the total value of his account at the end of each year. Use the following formula:

   $I = PRT$
   I = interest earned
   P = principle
   R = interest rate
   T = time

2. If a player purchases a car for $175,000 and the state sales tax rate is 8.5%, how much tax will he owe? What will be the total cost of the car?

3. If the value of the automobile in problem 2 depreciates by 5% each year, what will the car be worth at the end of two years? Construct a spreadsheet showing the amount of depreciation and the corresponding value of the car each year.

4. If a player purchases a house for $6,850,000 and the price of the home appreciates 15% a year for the next two years, what will be the value of the home at the end of that period? Construct a spreadsheet showing the amount of annual appreciation and the corresponding value of the house at the end of each year.

Number sense

QUIZ 23

Prime Factorization

1. The weekly point totals (in 42nds) for various players are listed below. Write the prime factorization of each number, using exponents.

Week	Point Totals
Week 1	48
Week 2	144
Week 3	17
Week 4	206
Week 5	81
Week 6	150
Week 7	57

2. List the first five prime numbers: ____ ____ ____ ____ ____

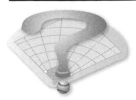

QUIZ 24

Scientific Notation

The rectangular dimensions of a spring training facility are 1200 feet by 1300 feet. Write the area in scientific notation for the following units of measurement.

1. Square inches:

2. Square yards:

3. Square centimeters:

4. Square millimeters:

5. Square meters:

Write the following in scientific notation.

6. 119.73

7. .0005

8. .100300

9. 167,822.10

10. 1.0

Write the following in standard form.

11. 777.774×10^{-4}

12. $91\frac{1}{8} \times 10^{5}$

QUIZ 25

Ordering Integers, Fractions, and Decimals

1. The following integers represent January temperatures for several cities that host baseball teams. Place them in ascending order on the number line below.

$$-5 \quad -12 \quad 71 \quad 54 \quad -20 \quad 42 \quad 61 \quad -6 \quad -4 \quad -1 \quad 66$$

2. The following integers represent the points earned for a player for the first ten games of the season. Place them in ascending order on the number line below.

$$-19 \quad 31 \quad -17 \quad -3 \quad 29 \quad -15 \quad 42 \quad -21 \quad 52 \quad 22$$

3. Place the following numerical values in ascending order on the number line below.

$$-.00999 \quad 12\frac{1}{2} \quad -11\frac{3}{7} \quad -99.009 \quad 3.8 \quad -20\frac{1}{6}$$

4. On the number line below, place the following point totals in ascending order.

$$\frac{5}{6} \quad \frac{37}{42} \quad \frac{2}{3} \quad .00775 \quad \frac{5}{7} \quad .070075 \quad .00575$$

Number sense

Operations with Integers

1. If a player hit 429 home runs in 11 seasons, how many home runs did he average per season?

2. The numerical values below represent the points earned by 10 players on a team. How many points did the team earn or lose?

   −82   11   −42   175   −110   34   −17   −25   −44   62

3. If a player earned −47 points for the first half of the season, how many points is he projected to lose for the whole season?

4. If a player pitched 224 innings in 32 games, how many innings per game did he pitch, on average?

5. The numbers below represent profit or loss for five teams for one year. What is the average profit or loss?

   −$1,976,005 $−3,111,657   −$2,722,311 $763,798 $1,776,400

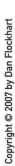

Number sense

Operations with Integers *(Cont'd.)*

6. If one team lost $5,978,800 while another team reported a profit of $2,656,432, how much greater was the second team's profit compared with the first team's?

7. If one team reported a loss of $10,111,250, which included a profit of $7,777,234 on parking fees, how much money did they lose on operations other than the parking fees?

Number sense

QUIZ 27

Permutations and Combinations

1. There are eight pitchers on a team. If the coach starts four pitchers during a four-game home stand, how many combinations can he choose from?

2. If a team has jerseys in two different styles, caps in four different styles, and pants in three different styles, how many combinations of uniforms do they have to choose from?

3. A team's uniform consists of four colors, but they have six colors to choose from. How many combinations of uniforms do they have?

4. Before a game, six umpires line up in single file for the National Anthem. In how many ways can the umpires line up?

QUIZ 28

Unit Conversions

1. The distance between bases on a baseball diamond is 30 yards. What is half of this distance in inches? In centimeters? (2.5 centimeters = 1 inch)

2. If a player stole 57 bases in one year, how many feet did he run to steal those bases?

3. The distance from home plate to the left field foul post is 365 feet. What is this distance in yards? In inches?

4. The distance from home plate to the rubber on the pitcher's mound is 60' 6". What is half of this distance in centimeters? In millimeters? *Hint:* 10 mm = 1 cm

5. A team spent 1,955 minutes practicing last week. How many hours did they spend practicing last week?

6. A team is scheduled to play their next game in exactly 3 days, 7 hours. How many hours will pass before they play their next game? How many minutes?

Algebra and functions

QUIZ 29

Evaluating Algebraic Expressions

Evaluate $\frac{1}{2}(H) + \frac{1}{3}(R + I) + \frac{1}{6}(B) + \frac{1}{7}(S + W) - \frac{1}{21}(K + E)$ if

1. $H = 3$
   $R = 4$
   $I = 5$
   $B = 3$
   $S = 2$
   $W = 1$
   $K = 1$
   $E = 2$

2. $H = 1$
   $R = 3$
   $I = 3$
   $B = 2$
   $S = 0$
   $W = 3$
   $K = 2$
   $E = 1$

Evaluate $\frac{1}{2}(V) + \frac{1}{3}(P) + \frac{1}{6}(K) - \frac{1}{7}(R) - \frac{1}{21}(B + W)$ if

3. $V = 1$
   $P = 7$
   $K = 11$
   $R = 6$
   $B = 12$
   $W = 4$

Algebra and functions

QUIZ 30

Properties of Mathematics

Use numerical values to write one equation for each of the properties listed.

Distributive property

Commutative property of addition

Commutative property of multiplication

Associative property of addition

Associative property of multiplication

Inverse property of addition

Inverse property of multiplication

Identity property of addition

Identity property of multiplication

Algebra and functions

QUIZ 31

Graphing on a Number Line

Use a number line to graph the range of points earned by your players in weeks 1–3.

1B   _____

2B   _____

3B   _____

SS   _____

P   _____

C   _____

DH   _____

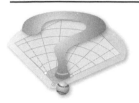

QUIZ 32

Linear Equations (A)

The equations below are used to compute total weekly points or to compute central angles in a circle graph. In each case, solve for the variable.

1. $\dfrac{1}{2}(1) + \dfrac{1}{3}(5 + I) + \dfrac{1}{6}(5) + \dfrac{1}{7}(1 + 0) - \dfrac{1}{21}(1 + 0) = 4\dfrac{2}{21}$

2. $\dfrac{1}{6}(0) + \dfrac{1}{9}(0 + 0) + \dfrac{1}{12}(0) + \dfrac{1}{18}(0 + 1) - \dfrac{1}{36}(K + 2) = -\dfrac{1}{9}$

3. $\dfrac{1}{9}(H) + \dfrac{1}{13.5}(3 + 2) + \dfrac{1}{18}(3) + \dfrac{1}{27}(3 + 2) - \dfrac{1}{54}(0 + 1) = \dfrac{25}{27}$

4. $\dfrac{1}{10}(1) + \dfrac{1}{15}(4 + 4) + \dfrac{1}{20}(B) + \dfrac{1}{30}(1 + 3) - \dfrac{1}{60}(0 + 3) = \dfrac{49}{60}$

Algebra and functions

Linear Equations (A) *(Cont'd.)*

5. $w \div 8 \times 360 = 180$

6. $5^4 (1) + 5^3 (9) + 5^2 (K) - 5^1 (1) - 5^0 (2 + 3) = 2015$

7. $\dfrac{1}{10} \div s \times 360 = 30$

8. $.4^0 (3) + .4^{-1} (1 + 2) + .4^{-2} (B) + .4^{-3} (0 + 2) - .4^{-4} (0 + 0) = 66.75$

9. $.2^{-3} (0) + \sqrt[3]{64} (3 + 0) + \left(\dfrac{3}{4}\right)^{-2} (B) + .025 (4 + 2) - \left(\dfrac{2}{8}\right)^1 (2 + 1) = 11.4$

Algebra and functions

153

Linear Equations (B)

In the problems below, insert the variables into the total points equation. Then solve for T and write the answer in its simplest form.

$$\frac{1}{2}(H) + \frac{1}{3}(R + I) + \frac{1}{6}(B) + \frac{1}{7}(S + W) - \frac{1}{21}(K + E) = T$$

1. $H = 2$
   $R = 3$
   $I = 6$
   $B = 3$
   $S = 1$
   $W = 1$
   $K = 1$
   $E = 1$

2. $H = 1$
   $R = 2$
   $I = 3$
   $B = 1$
   $S = 2$
   $W = 3$
   $K = 2$
   $E = 1$

3. $H = 0$
   $R = 1$
   $I = 0$
   $B = 2$
   $S = 4$
   $W = 2$
   $K = 2$
   $E = 2$

Algebra and functions

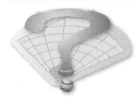

QUIZ 34

Area of Rectangles

1. Explain the meaning of the variables in the following equations:

$$P = 2l + 2w$$

$$A = bh$$

2. The dimensions of a rectangular stadium are 875 feet by 875 feet. The dimensions of the rectangular parking lot are 925 feet by 925 feet. Complete the table below. Do you see any patterns? Explain.

	Area of Stadium	Area of Parking Lot	Ratio of Area of Stadium to Area of Parking Lot
Square feet			
Square inches			
Square yards			
Square centimeters (2.5 cm = 1 inch)			
Square millimeters			

3. How much would it cost to resurface a baseball field if artificial turf costs $80 per square foot and the area that needs to be covered is 85,000 square feet?

QUIZ 35

Golden Rectangles

1. The ratio of length to width in a Golden Rectangle is approximately 1.6:1. Fill in the chart below.

Playing Area	Dimensions: $\frac{1}{2}$ Length by $\frac{1}{2}$ Width	Ratio of $\frac{1}{2}$ Length to $\frac{1}{2}$ Width	Difference from Golden Rectangle Ratio
Clubhouse	45 ft. by 19 ft.		
Dugout	30 ft. by 3 ft.		
Basketball court	35 ft. by 21 ft.		
Soccer field	40 meters by 27.5 meters		

2. Which player areas have a ratio that approximates that of a Golden Rectangle?

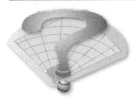

QUIZ 36

Functions

In the exercises below, write the function rule and solve for the variable.

1. X = number of runs batted in; Y = points earned.

Function rule: _____

X	Y
3	1
9	3
27	9
81	n

2. X = number of bases on balls; Y = points earned.

Function rule: _____

X	Y
4	$\frac{4}{7}$
6	$\frac{6}{7}$
8	$1\frac{1}{7}$
10	n

Measurement and geometry

Functions *(Cont'd.)*

3. X = number of errors; Y = number of points earned.

   Function rule: _____

X	Y
5	$\frac{5}{21}$
10	n
15	$\frac{5}{7}$
20	$\frac{20}{21}$

4. Below, construct your own function chart.

   Function rule: _____

X	Y

Measurement and geometry

QUIZ 37

Area and Circumference of Circles

1. A circular logo is located at the center of a baseball field and has a diameter of 30 feet. Find the area and circumference of the logo.

2. If the area of the on-deck circle is 19.625 square feet, what is the diameter of the circle?

3. An on-deck circle has a diameter of 4 feet. What is the area of the circle?

4. A circular baseball stadium has a radius of 950 feet. Find the diameter, circumference, and area of the stadium.

Diameter: _____

Circumference: _____

Area: _____

5. If the circumference of a logo on a shirt is 2.25 inches, what are the radius, diameter, and area of the logo?

Radius: _____

Diameter: _____

Area: _____

Measurement and geometry

QUIZ 38

Weight

1. Answer the following questions based on the table below.

	Actual Weight		
	Pounds	**Ounces**	**Grams**
Football		14	14
Basketball	1	5	5
Baseball		5	3
Soccer ball		15	4
Table tennis ball			2.5
Hockey puck		6	

For each of the following problems, predict the answer, then solve the problem.

2. How many hockey pucks would it take to weigh as much as ten baseballs? As much as 300 table tennis balls?

3. Which is greater, the weight of six hockey pucks or five basketballs?

4. Which is less, the weight of 6,000 table tennis balls or 30 baseballs?

Measurement and geometry

QUIZ 39

The Pythagorean Theorem

Use the Pythagorean Theorem to solve the following problems:

1. The distance between consecutive bases on a baseball diamond is 30 yards. Find the distance from first base to third base.

2. Find the length of the diagonal of a parking lot if the length of the lot is 2,500 feet and the width is 1,900 feet.

3. Find the length of the diagonal of a clubhouse if the length of the clubhouse is 90 feet and the width is 40 feet.

4. Find the width of a dugout if the dugout's length is 16 meters and the length of the diagonal is 18 meters.

5. Find the distance between the two foul poles at a stadium if both are 360 feet from home plate.

Measurement and geometry

Name _____

Mean, Median, Mode, Range

(Use with Handout 12)

For each of the first three weeks, find the points earned by each player on the English Bulldogs. In the table below, use fractions to record the mean, median, mode, and range of those points earned for each of the first six weeks.

Week	Mean	Median	Mode	Range
1				
2				
3				

Statistics, data analysis, and probability

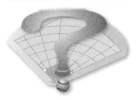

QUIZ 41
Probability

1. Last year, Antonio Jordan sprayed 40% of his hits to the left side of the field, 30% to the right side, and 30% straightaway. If Jordan had 220 hits last year, how many hits did he have in each direction?

2. Using only the data in problem 1, find the probability that Jordan's first hit this year will be to the right side of the field.

3. A team's record for the last eight years is 584–712. Without taking any other variables into account, what should the team's record be this year?

4. Express numerically in several ways that the probability of an event occurring is 75%.

5. If the probability of an event occurring is 23%, what is the probability of that event not occurring?

Probability *(Cont'd.)*

6. The letters in "Sy Bojaan" are placed in a hat. Find the probability of the following random events:

   A. Selecting the letter *j*

   B. Selecting the letters *s, o,* or *n*

   C. Selecting the letter *a*

   D. Selecting any letter except *a*

   E. Selecting the letter *y,* replacing it, then selecting the letter *y* again

   F. Selecting the letters *o* and *j* on consecutive draws (without replacing letters)

In exercises 7–11, you are given $P(Q)$, the probability that a player will hit a home run at various ballparks. Find $P(\text{not } Q)$, the probability that event Q will not occur.

7. $P(Q) = \dfrac{5}{21}$ $P(\text{not } Q) =$

8. $P(Q) = .204$ $P(\text{not } Q) =$

9. $P(Q) = 17\%$ $P(\text{not } Q) =$

10. $P(Q) = 1$ $P(\text{not } Q) =$

11. $P(Q) = 0$ $P(\text{not } Q) =$

Statistics, data analysis, and probability

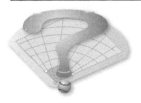

QUIZ 42
Circle Graphs

(Use with Handout 12)

1. Find the central angles for Felipe Vargas, Louie Carter, and Julio Martinez for week 1.

2. Find the central angles for Felipe Vargas, Louie Carter, and Julio Martinez for week 4.

3. If the central angle in a circle graph is 225 degrees, what percentage of the graph will that section represent?

4. If the central angle in a circle graph is 5 degrees, what percentage of the graph will that section represent?

5. If one section of a circle graph represents 43% of the total graph, what is the measurement of the corresponding central angle?

6. The sum of the percentages inside a circle graph is 359 degrees. Explain how this could occur.

Stem-and-Leaf Plots
and Histograms

1. The following values represent the point totals (in 42nds) for the English Bulldogs for the last weeks of the season. Using graph paper, construct a stem-and-leaf plot and a histogram of the data.

71	12	45	89	82	56	51
0	41	85	44	73	49	

2. The following values represent the point totals (in 42nds) for Stanley Bryant for the last weeks of the season. Using graph paper, construct a stem-and-leaf plot and a histogram of the data.

31	57	37	0	71	11	47	84
101	96	0	29	4	0	16	

Statistics, data analysis, and probability

QUIZ 44

Scatter Plots

1. The table below represents the number of home runs and runs batted in by a player over the course of several seasons. On graph paper, construct a scatter plot of this data. Does the scatter plot have a positive or negative correlation? Explain.

Home Runs	Runs Batted In
49	137
32	105
29	88
33	99
34	114
46	120
41	107
15	67
9	43
25	71

2. The table below shows the number of innings pitched and strikeouts for several pitchers for one season. Using graph paper, construct a scatter plot of this data. Does the scatter plot have a positive or negative correlation? Explain.

Player	Number of Innings Pitched	Number of Strikeouts
A	229	204
B	171	165
C	166	101
D	34	37
E	144	56
F	127	104
G	67	58
H	21	30
I	195	188

Statistics, data analysis, and probability 167

QUIZ 45

Box-and-Whisker Plots

The following data sets represent the points earned (in 42nds) by Leonard Hamilton and the total team points earned by the English Bulldogs for twelve weeks of the season. Draw a box-and-whisker plot for each set of data. On each plot, label the median as well as the upper and lower quartiles.

1. Leonard Hamilton   3, 40, 57, 0, 112, 11, 0, 32, 77, 4, 0, 39

2. The English Bulldogs   301, 159, 68, 12, 17, 4, 116, 171, 49, 43, 74, 83

Statistics, data analysis, and probability

Statements Using Math Terminology

(Use with Handout 12)

Write five statements, and show the mathematical proof for each.

1.

2.

3.

4.

5.

Chapter Six

Assessment

The purpose of assessment is to gather evidence of student learning. The pretest/posttest in this section can be used to accomplish that task because it represents a compilation of the concepts used on all of the practice worksheets. That said, the pretest/posttest covers a wide range of concepts, some of which may be too advanced for younger students. If that is the case, you can assign specific problems on the pretest/posttest to meet the needs of your students.

The pretest/posttest can be used as diagnostic and summative assessments. Diagnostic assessment occurs at the beginning of a unit and provides teachers with an understanding of the skills, knowledge, and learning needs students bring to the learning environment. Summative assessment occurs at the end of a period of learning and provides students with opportunities to demonstrate their achievement of the learning objectives.

Used diagnostically, giving the pretest/posttest prior to playing the Fantasy game can help you ascertain the mathematical concepts in which students need reinforcement as well as the areas in which they are strong. For example, if results from the pretest/posttest indicate that students do not yet have the skills to convert between fractions and decimals, then you probably would not use a scoring system that includes both fractions and decimals. As another example, if the pretest/posttest indicated students need more reinforcement rounding decimals, then you could integrate Practice Worksheet 7 and Quiz 7 (Rounding Decimals) into your Fantasy game.

You have the option to give the pretest/posttest at the end of the Fantasy game so you can record student achievement and growth over time as it relates to the content of the unit. Comparing the results of these two tests can yield empirical data in terms of the concepts students need to improve on and areas of strength. Data from the results given after playing the Fantasy game could also be used to modify the game in the future to better meet the needs of students.

Name _____

Pretest/Posttest

Show all of your work.

1. Find the sum of the points earned by the following players:

Ryan King	$\dfrac{36}{42}$	Bruce Reed	$\dfrac{1}{3}$
Scott Green	$\dfrac{5}{6}$	Jonathan Cook	$\dfrac{5}{21}$
Tre Collins	$\dfrac{3}{7}$	Willie Bell	$\dfrac{17}{42}$
Lou Stewart	$\dfrac{1}{2}$	Ray Mitchell	$\dfrac{5}{14}$

2. In problem 1, what is the ratio of the points earned by Reed to the points earned by Stewart?

3. In problem 1, convert King's points to a decimal and round to the nearest thousandth.

4. Evaluate

$$\frac{1}{2}(H) + \frac{1}{3}(R + I) + \frac{1}{6}(B) + \frac{1}{7}(S + W) - \frac{1}{21}(K + E)$$

when

$H = 3$
$R = 2$
$I = 5$
$B = 3$
$S = 2$
$W = 2$
$K = 1$
$E = 1$

Pretest/Posttest *(Cont'd.)*

5. If one factor of $\frac{24}{48}$ is $\frac{8}{12}$, what is the second factor?

6. Write the prime factorization of 180, using exponents.

7. Convert $\frac{87}{42}$ into a mixed number, and write it in the simplest form.

8. Which is the higher average per game: 78 hits in 91 games or 95 hits in 130 games?

9. If Gerald Cox accumulated 13.5 points during the first 27 games of the season, how many points is he projected to earn for an entire season (162 games)?

Assessment

Pretest/Posttest *(Cont'd.)*

10. Based on the data in problem 1, find the following:

Range:

Mean:

Median:

Mode:

11. Fill in the missing numbers in the patterns below.

   A. Ralph Bailey $\dfrac{1}{24}$ $\dfrac{3}{48}$ $\dfrac{1}{12}$ _____

   B. Andrew Edwards $\dfrac{1}{16}$ $\dfrac{1}{8}$ $\dfrac{9}{48}$ _____

12. The price of an autographed jersey rose from \$125 to \$235. Find the percentage of price increase.

13. If a player invests 25% of his annual salary of \$9.1 million at 6.5%, how much interest will he earn after one year?

14. The dimensions of the parking lot are 1000 feet by 2220 feet. Find the area of the parking lot in square inches.

15. In problem 14, what is the length of the diagonal of the parking lot?

Pretest/Posttest *(Cont'd.)*

16. The letters in "Rapar Garcia" are placed in a hat. Find the probability of the following random events:

    A. Selecting the letter *r*

    B. Selecting the letters *p, i,* or *c*

17. Solve for the variable in the following equation:

$$\frac{1}{10}(2) + \frac{1}{15}(R + 7) + \frac{1}{20}(4) + \frac{1}{30}(0 + 1) - \frac{1}{60}(1 + 2) = 1\frac{1}{20}$$

Assessment

Chapter Seven

Answer Keys

Practice Worksheet 1

1. | *Nearest $10,000* | *Nearest $100,000* | *Nearest $1,000,000* |
    | --- | --- | --- |
    | $6,780,000 | $6,800,000 | $7,000,000 |
    | $5,000,000 | $5,000,000 | $5,000,000 |
    | $3,500,000 | $3,500,000 | $4,000,000 |
    | $2,990,000 | $3,000,000 | $3,000,000 |

2. $6,444,700 = $6,000,000 + $400,000 + $40,000 + $4,000 + $700

    $7,200,066 = $7,000,000 + $200,000 + $60 + $6

    $4,950,222 = $4,000,000 + $900,000 + $50,000 + $200 + $20 + $2

    $1,033,003 = $1,000,000 + $30,000 + $3,000 + $3

    $5,999,999 = $5,000,000 + $900,000 + $90,000 + $9,000 + $900 + $90 + $9

Practice Worksheet 2

Player B: 72, 4   Player C: 84, 6   Player D: 120, 5   Player E: 90, 3

Practice Worksheet 3

1. $34,349,945   2. $127,983,025   3. $25,596,605   4. 3   5. $59,500,000

Practice Worksheet 4

Brady: $\dfrac{8}{14}$ $\dfrac{12}{21}$ $\dfrac{16}{28}$ Martinez: $\dfrac{26}{84}$ $\dfrac{39}{126}$ $\dfrac{52}{168}$ Carter: $\dfrac{6}{8}$ $\dfrac{9}{12}$ $\dfrac{12}{16}$ Blanco: $\dfrac{10}{12}$ $\dfrac{15}{18}$ $\dfrac{20}{24}$

Flores: $\dfrac{10}{48}$ $\dfrac{15}{72}$ $\dfrac{20}{96}$ Cooke: $\dfrac{22}{42}$ $\dfrac{33}{63}$ $\dfrac{44}{84}$

Practice Worksheet 5

1. $\dfrac{1}{3}$ $\dfrac{2}{3}$ 1   2. $\dfrac{1}{6}$ $\dfrac{1}{3}$ $\dfrac{1}{2}$ 3. $\dfrac{13}{21}$ $1\dfrac{5}{21}$ $1\dfrac{6}{7}$ 4. $\dfrac{5}{21}$ 5. 1.5

Practice Worksheet 6

1. $1\dfrac{13}{14} > 1\dfrac{2}{21} > 1\dfrac{1}{42} > \dfrac{13}{21} > \dfrac{25}{42} > \dfrac{3}{7} > \dfrac{1}{3}$

2. .333 < .429 < .595 < .619 < 1.024 < 1.095 < 1.929

Practice Worksheet 7

Table 1:   Table 2: Answers will vary.

Carter:	2.5	2.52	2.524
Martinez:	5.7	5.74	5.738
Vargas:	3.1	3.10	3.095

Practice Worksheet 8

1. $2\frac{31}{42}$ 2. $6\frac{2}{3}$ 3. $3\frac{29}{42}$ 4. $\frac{13}{14}$ 5. $4\frac{31}{42}$ 6. $\frac{42}{115}$ 7. $\frac{3}{20}$ 8. $\frac{42}{155}$ 9. $\frac{14}{13}$ 10. $\frac{42}{199}$

Practice Worksheet 9

1. $3\frac{10}{21}$ 2. $\frac{29}{42}$ 3. $7\frac{1}{6}$ 4. $3\frac{3}{14}$ 5. $2\frac{9}{14}$

Practice Worksheet 10

1.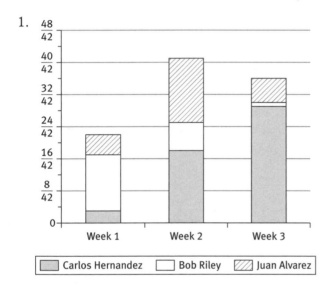

Practice Worksheet 11

1. 9   2. $\frac{2}{9}$ 3. $\frac{1}{2}$ 4. $\frac{3}{4}$ 5. 324

Practice Worksheet 12

Table 1:

Table 2: Answers will vary.

Vargas:	3	$3\frac{1}{6}$	$3\frac{1}{7}$
Carter:	$2\frac{1}{2}$	$2\frac{3}{6}$	$2\frac{4}{7}$
Martinez:	$5\frac{1}{2}$	$5\frac{4}{6}$	$5\frac{5}{7}$

Practice Worksheet 13

1. 81.30   2. $16,285.71 per day; $2,035.71 per hour; $33.93 per minute; $0.57 per second
3. first to second base: 12.5 hours; 1 mile: 733.33 hours   4. 1,424,250 oz.   5. $130.51

Practice Worksheet 14

1. 17.5 cents per ounce (20-ounce size)   2. Richard: $26.\overline{6}$ miles per gallon; Hayes: $25.\overline{3}$ miles per gallon   3. $300,000 per acre (for the 15 acres)   4. $14.81 (season ticket)   5. $14.5 million for 15 years provides the higher annual salary: $966,666.67

Practice Worksheet 15

1. Carter:	$2\frac{11}{21}$	2.5238	2.5	2.52	2.524	252.4%
Martinez:	$5\frac{31}{42}$	5.7381	5.7	5.74	5.738	573.8%

2. Answers will vary.

Practice Worksheet 16

1. 324.0%   2. 96.2%   3. 53.9%   4. 536.3%   5. 18.6%

Practice Worksheet 17

1. 22.7%   2. 10.9%   3. 4.9%   4. 7.8%   5. 11.1%

Practice Worksheet 18

1. 13.1%   2. 225.7%   3. 153.2%   4. $10\frac{2}{3}$ 5. 1   6. $\frac{1}{2}$ 7. $4\frac{3}{4}$

8.

	Bats fans in Chicago	**Rats fans in Boston**
After year 1:	450,000	437,000
After year 2:	405,000	415,150
After year 3:	364,500	394,392
After year 4:	328,050	374,672

Practice Worksheet 19

1. $42\frac{5}{8}$ 2. 7   3. 3   4. 6   5. 7.5   6. 224   7. $227\frac{1}{3}$ 8. 8.375 inches   9. 60 ft.   10. 44.55 hours

Practice Worksheet 20

1. $\frac{2}{3}$ 2. $1\frac{5}{7}$ or 1.714   3. 6.875 or $6\frac{7}{8}$ 4. 3.3 or $3\frac{3}{10}$ 5. 6.46

Practice Worksheet 21

1. $\frac{4}{7}$ 2. $\frac{2}{3}$, 5   3. 400 ft. by 500 ft.   4. $\frac{13}{6}$ 5. $\frac{3}{56}$

Practice Worksheet 22

1.
Year	Interest Earned	Account Value
1	$196,875	$3,946,875
2	$207,210.94	$4,154,085.94

2. tax: $15,675; total cost: $205,675

3.
Year	Amount of Depreciation	Value of Car
1	$19,000	$171,000
2	$17,100	$153,900

4.
Year	Amount of Appreciation	Value of House
1	$425,000	$4,675,000
2	$467,500	$5,142,500

Practice Worksheet 23

1. week 2: $3^2 \times 11$; week 3: 7; week 4: 2^6; week 5: 5^3; week 6: 0; week 7: 19

2. 2, 3, 5, 7, 11

Practice Worksheet 24

1. 2.376×10^8 2. 1.83×10^5 3. 1.485×10^9 4. 1.485×10^{11} 5. 1.485×10^5

6. 7.227×10^1 7. 8×10^{-6} 8. 9.00002×10^{-1} 9. 1.42887511×10^5 10. 7.10605008×10^7

11. .33005   12. 16,750

Practice Worksheet 25

1. $-31, -14, -4, -2, -1, 22, 34, 41, 61, 65, 86$

2. $-51, -39, -38, -25, -13, -3, 22, 37, 71, 91$

3. $-4\frac{5}{16}, -3.888, -2\frac{7}{8}, -.02258, 6.004, 8\frac{2}{5}$ 4. $.00235, .0235, \frac{1}{6}, .235, \frac{3}{7}, \frac{13}{21}, \frac{5}{7}$

Practice Worksheet 26

1. 42   2. lost 20 points   3. 52   4. 6   5. $4,440,241.40 average profit   6. $16,635,302

7. $5,888,464

Practice Worksheet 27

1. $\dfrac{7!}{3!(7-3)!} = \dfrac{7!}{3!4!} = \dfrac{7 \times 6 \times 5 \times 4 \times 3 \times 2 \times 1}{3 \times 2 \times 1 \times 4 \times 3 \times 2 \times 1} = 35$ 2. 18   3. 10   4. 120

Practice Worksheet 28

1. 1,080 inches; 2,700 cm   2. 4,050   3. $128.\overline{3}$ yards; 4,620 inches   4. 1815 cm; 18150 mm
5. 46.08 hours or 46 hours, 5 minutes   6. 62 hours; 3,720 minutes

Practice Worksheet 29

1. $4\dfrac{5}{42}$ 2. $3\dfrac{1}{2}$ 3. $3\dfrac{4}{21}$

Practice Worksheet 30

Answers will vary.

Practice Worksheet 31

Answers will vary.

Practice Worksheet 32

1. $H = 2$ 2. $I = 3$ 3. $R = 1$ 4. $B = 4$ 5. $w = 5$ 6. $P = 7$ 7. $s = 3$ 8. $K = 2$
9. $w = 1.125$ 10. $W = 2$ 11. $E = 3$ 12. $S = 1$ 13. $R = 3$ 14. $V = 2$ 15. $W = 22$
16. $P = 18$ 17. $K = -3$

Practice Worksheet 33

1. $2\dfrac{4}{7}$ 2. $2\dfrac{1}{2}$ 3. $4\dfrac{5}{6}$

Practice Worksheet 34

1. Perimeter = 2 × length + 2 × width; Area = base × height

2.

	Area of Ballpark	*Area of Store*	*Ratio*
Square feet	810,000	902,500	.898
Square inches	116,640,000	129,960,000	.898
Square yards	90,000	100,277.8	.898
Square centimeters	729,000,000	812,250,000	.898
Square millimeters	72,900,000,000	81,225,000,000	.898
Square meters	72,900	81,225	.898

3. 6,000,000   4. Statements 1 and 2: Answers will vary.

Area: Dugout: 360 sq. ft.; soccer field: 4,400 square meters; basketball court: 3,825 sq. ft.; clubhouse: 3,420 sq. ft. *Perimeter:* Dugout: 132 ft.; soccer field: 270 m.; basketball court: 260 ft.; clubhouse: 256 ft.

Practice Worksheet 35

1. Clubhouse: 2.4:1   .8
   Dugout: 10:1   8.4
   Basketball: 1.7:1   .1
   Soccer field: 1.5:1   .1

2. basketball court and soccer field   3. Answers will vary.   4. Answers will vary.

Practice Worksheet 36

1. $\frac{1}{2}(X) = Y$; $n = 12$ 2. $\frac{1}{3}(X) = Y$; $n = 5\frac{1}{3}$ 3. $\frac{1}{7}(X) = Y$; $n = 5$ 4. Answers will vary.

Practice Worksheet 37

1. area: 660.185 sq. ft.; circumference: 91.06 ft.   2. 6 ft.   3. 7.065 sq. ft.

4. diameter: 1,600 ft.; circumference: 5,024 ft.; area: 2,009,600 sq. ft.

5. radius: .56 inches; diameter: 1.115 inches; area: .98 sq. inches

Practice Worksheet 38

Answers to items 1–4 will vary.

Practice Worksheet 39

1. 42.43 yards　　2. 3,498.57 ft.　　3. 116.62 ft.　　4. 5.74 meters　　5. 487.90 ft.

Practice Worksheet 40

1.

	Mean	Median	Mode	Range
Week 4	$\frac{19}{42}$	$\frac{1}{6}$	none	$1\frac{4}{21}$
Week 5	$\frac{97}{126}$	$\frac{41}{42}$	none	$1\frac{1}{3}$
Week 6	$\frac{5}{9}$	$\frac{11}{42}$	none	$1\frac{17}{42}$

2. Answers will vary.

Practice Worksheet 41

1. left: 105; center: 63; right: 42　　2. 50%　　3. 90–72　　4. .33, $\frac{33}{100}$, $\frac{1}{3}$; other answers are possible.　　5. 85%

6. A. $\frac{1}{12}$　B. $\frac{1}{4}$　C. $\frac{1}{4}$　D. $\frac{11}{12}$　E. $\frac{1}{144}$　F. $\frac{1}{66}$　7. $\frac{28}{42} = \frac{2}{3}$　8. .605　9. 73%　10. 0　11. 1

Practice Worksheet 42

1. Vargas: 0°; Carter: 208°; Martinez: 152°　　2. Vargas: 174°; Carter: 0°; Martinez 186°

3. Vargas 98°; Carter, 80°; Martinez, 182°　　4. 19.4%　　5. 40.3%　　6. 223.2°

Practice Worksheet 43

1. stem-and-leaf plot:

2 | 3
3 | 4, 8
4 | 9
5 | 2
6 | 1, 6
7 | 3, 7
8 | 5, 8
9 | 8

histogram:

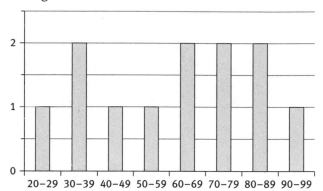

2. stem-and-leaf plot:

0 | 0, 0, 0, 3, 4, 8
1 | 0, 2, 6, 7
2 | 9
3 | 7, 7
6 | 6
7 | 1
9 | 0

histogram:

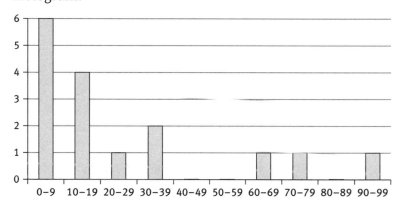

Practice Worksheet 44

1. Negative correlation is a statistical relationship in which an increase in the value of one variable is accompanied by a decrease in the value of the other variable.

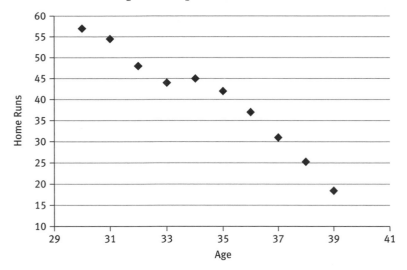

2. Positive correlation is a statistical relationship in which an increase in the value of one variable is accompanied by an increase in the value of the other variable.

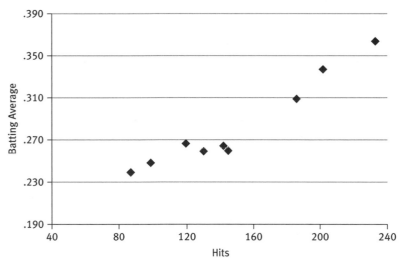

Practice Worksheet 45

1.

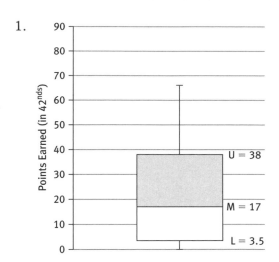

2.

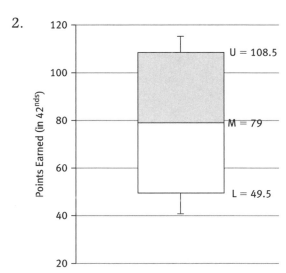

Practice Worksheet 46

1. $5\frac{31}{42} > 3\frac{4}{42} + 2\frac{22}{42}$ 2. $\left(\frac{25}{42} + 1\frac{13}{14}\right) \div 3\frac{1}{7} = .804 = 80.4\%$ 3. $1\frac{4}{21} \div 1\frac{15}{42} = .87719 = 88\%$

4. $1\frac{1}{2} - 1\frac{19}{42} = \frac{2}{42} = \frac{1}{21}$ 5. Answers to items A–E will vary.

Extra Credit Problems

1. 16   2. Answers to items A–D will vary.   3. Answers will vary.   4. Answers will vary.

Quiz 1

1.

Nearest $10,000	Nearest $100,000	Nearest $1,000,000
$3,330,000	$3,300,000	$3,000,000
$1,910,000	$1,900,000	$2,000,000
$3,560,000	$3,600,000	$4,000,000
$5,670,000	$5,700,000	$6,000,000
$4,460,000	$4,500,000	$4,000,000

2. $4,001,001 = $4,000,000 + $1,000 + $1

   $6,449,501 = $6,000,000 + $400,000 + $40,000 + $9,000 + $500 + $1

   $3,888,888 = $3,000,000 + $800,000 + $80,000 + $8,000 + $800 + $80 + $8

   $3,100,704 = $3,000,000 + $100,000 + $700 + $4

   $9,999,999 = $9,000,000 + $900,000 + $90,000 + $9,000 + $900 + $90 + $9

Quiz 2

Player A: 105, 35; Player B: 36, 1; Player C: 87, 29; Player D: 42, 2; Player E: 56, 1

Quiz 3

1. $37,000,635   2. $129,633,105   3. $25,926,621   4. 1.5   5. $55,000,000

Quiz 4

1. $\dfrac{6}{8}$ $\dfrac{9}{12}$ $\dfrac{12}{16}$ 2. $\dfrac{6}{14}$ $\dfrac{9}{21}$ $\dfrac{12}{28}$ 3. $\dfrac{34}{42}$ $\dfrac{51}{63}$ $\dfrac{68}{84}$ 4. $\dfrac{16}{24}$ $\dfrac{24}{36}$ $\dfrac{32}{48}$ 5. $\dfrac{16}{42}$ $\dfrac{24}{63}$ $\dfrac{32}{84}$

6. $\dfrac{6}{16}$ $\dfrac{9}{24}$ $\dfrac{12}{32}$ 7. $\dfrac{10}{12}$ $\dfrac{15}{18}$ $\dfrac{20}{24}$

Quiz 5

1. $\dfrac{41}{42}$ $1\dfrac{20}{21}$ $2\dfrac{13}{14}$ 2. $\dfrac{3}{7}$ $\dfrac{6}{7}$ $1\dfrac{2}{7}$ 3. $\dfrac{11}{42}$ $\dfrac{11}{21}$ $\dfrac{11}{14}$ 4. $\dfrac{3}{7}$ 5. 1.25

Quiz 6

1. $\dfrac{1}{6} < \dfrac{1}{3} < \dfrac{3}{7} < 1\dfrac{1}{42} < 1\dfrac{2}{21} < 1\dfrac{4}{21}$

2. $1.190 > 1.095 > 1.024 > .429 > .333 > .167$

3. $\dfrac{1}{6} < \dfrac{3}{7} < \dfrac{25}{42} < \dfrac{13}{21} < 1\dfrac{1}{42} < 1\dfrac{4}{21} < 1\dfrac{1}{3}$

4. $1.333 > 1.190 > 1.024 > .619 > .595 > .429 > .167$

Quiz 7

Table 1:   Table 2: Answers will vary.

Carter:	1.9	1.93	1.929
Martinez:	2.4	2.40	2.405
Vargas:	2.2	2.21	2.214

Quiz 8

1. $1\dfrac{1}{2}$ 2. $4\dfrac{2}{21}$ 3. 2 4. $2\dfrac{17}{42}$ 5. $5\dfrac{8}{21}$ 6. $\dfrac{2}{3}$ 7. $\dfrac{21}{86}$ 8. $\dfrac{1}{2}$ 9. $\dfrac{42}{101}$ 10. $\dfrac{21}{113}$

Quiz 9

1. $6\dfrac{1}{42}$ 2. $\dfrac{29}{42}$ 3. $6\dfrac{4}{21}$ 4. $3\dfrac{3}{14}$ 5. $2\dfrac{9}{14}$

Quiz 10

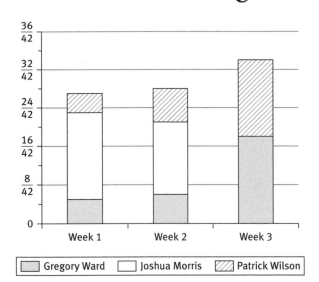

Quiz 11

1. 14   2. $\frac{4}{7}$ 3. $\frac{17}{21}$ 4. $\frac{3}{4}$ 5. 12

Quiz 12

Table 1:

Vargas:	2	$2\frac{1}{6}$	$2\frac{2}{7}$
Carter:	2	2	2
Martinez:	$2\frac{1}{2}$	$2\frac{2}{6}$	$2\frac{3}{7}$

Table 2: Answers will vary.

Quiz 13

1. 79.53   2. $18,000.00 per day; $2,250.00 per hour; $37.50 per minute; $0.63 per second
3. first to second base: 25 hours; 1 mile: 1466.67 hours   4. 2,747,250 oz.   5. $124.60

Quiz 14

1. 19 cents per ounce (18-ounce size)　2. Turner: 28.93 miles per gallon; Wright: 31.74 miles per gallon　3. $138,888.89 per acre (for the 45 acres)　4. $16.67 (season ticket)　5. $1,541,666.67 per year ($9.25 million for 6 years)

Quiz 15

1. Vargas:　$1\frac{9}{14}$　1.6429　1.6　1.64　1.643　164.3%　　2. Answers will vary.

Carter:　$1\frac{1}{42}$　1.0238　1.0　1.02　1.024　102.4%

Martinez:　$3\frac{5}{14}$　3.3571　3.4　3.36　3.357　335.7%

Quiz 16

1. 77.8%　2. undefined　3. 48.9%　4. 73.2%　5. 18.6%

Quiz 17

1. 22.2%　2. 8.6%　3. 7.9%　4. 6.3%　5. 8.6%

Quiz 18

1. 58.3%　2. 178.8%　3. 42.9%　4. $\frac{2}{3}$

Quiz 19

1. $3\frac{3}{4}$ or 3.75　2. 255　3. 140　4. $39\frac{13}{21}$ or 39.619　5. 10　6. 25

Quiz 20

1. $\frac{5}{6}$ or $.8\overline{3}$ 2. 0.75   3. $1\frac{3}{7}$ or 1.429   4. $5\frac{1}{3}$ or $5.\overline{3}$ 5. 4

Quiz 21

1. $\frac{6}{7}$ 2. Answers will vary; one possible answer is $\frac{8}{1}, \frac{3}{7}$ 3. 300 ft. by 600 ft.; 900 ft. by 200 ft.

4. $\frac{8}{3}$ 5. $\frac{6}{45}$

Quiz 22

1.

Year	Interest	Value of Account
1	$95,625	$2,345,625
2	$99,689.06	$2,445,314.06

2. tax: $14,875; total cost: $189,875

3.

Year	Amount of Depreciation	Value of Car
1	$8,750	$166,250
2	$8,312.50	$157,937.50

4.

Year	Amount of Appreciation	Value of House
1	$1,027,500	$7,877,500
2	$1,181,625	$9,059,125

Quiz 23

1. week 1: $2^4 \times 3$; week 2: $2^4 \times 3^2$; week 3: 17; week 4: 2×103; week 5: 3^4; week 6: $2 \times 3 \times 5^2$; week 7: 3×19 2. 2, 3, 5, 7, 11

Quiz 24

1. 2.2464×10^8 2. 1.7333×10^5 3. 1.404×10^9 4. 1.404×10^{11} 5. 1.404×10^5
6. 1.1973×10^2 7. 5.0×10^{-4} 8. 1.003×10^{-1} 9. 1.678221×10^5 10. 1.0×10^0
11. .0777774   12. 9,112,500

Quiz 25

1. -20, -12, -6, -5, -4, -1, 42, 54, 61, 66, 71
2. -21, -19, -17, -15, -3, 22, 29, 31, 42, 52
3. -99.009, $-20\frac{1}{6}$, $-11\frac{3}{7}$, $-.00999$, 3.8, $12\frac{1}{2}$ 4. .00575, .00775, .070075, $\frac{2}{3}$, $\frac{5}{7}$, $\frac{5}{6}$, $\frac{37}{42}$

Quiz 26

1. 39   2. lost 38 points   3. 94   4. 7   5. \$1,053,955 average loss   6. \$8,635,232
7. \$17,888,484

Quiz 27

1. 70   2. 24   3. 15   4. 720

Quiz 28

1. 540 inches; 1,350 cm   2. 5,130 ft.   3. $121.\overline{6}$ yds.; 4,380 inches   4. 907.5 cm; 9075 mm
5. 32.583 hrs.   6. 79 hrs.; 4,740 minutes

Quiz 29

1. $5\frac{2}{7}$ 2. $3\frac{5}{42}$ 3. $3\frac{1}{21}$

Quiz 30

Answers will vary.

Quiz 31

Answers will vary.

Quiz 32

1. $I = 3$ 2. $K = 4$ 3. $H = 2$ 4. $B = 2$ 5. $w = 4$ 6. $K = 11$ 7. $s = 1.2$ 8. $B = 4$ 9. $B = 0$

Quiz 33

1. $4\frac{29}{42}$ 2. $2\frac{19}{21}$ 3. $1\frac{1}{3}$

Quiz 34

1. Perimeter = 2 × length + 2 × width; Area = base × height

2.

	Area of Stadium	Area of Parking Lot	Ratio
Square feet	765,625	855,625	.895
Square inches	110,250,000	123,210,000	.895
Square yards	85,069.$\overline{4}$	95,069.$\overline{4}$	.895
Square centimeters	689,062,500	770,062,500	.895
Square millimeters	68,906,250,000	77,006,250,000	.895

3. $6,800,000

Quiz 35

1. Clubhouse: 2.4:1   .8
   Dugout: 10:1   8.4
   Basketball court: 1.7:1   .1
   Soccer field: 1.5:1   .1

2. Basketball court and soccer field

Quiz 36

1. $\frac{1}{3}(X) = Y$; $n = 27$ 2. $\frac{1}{7}(X) = Y$; $n = 1\frac{3}{7}$ 3. $\frac{1}{21}(X) = Y$; $n = \frac{10}{21}$ 4. Answers will vary.

Quiz 37

1. area: 706.5 sq. ft.; circumference: 99.42 ft.   2. 5 ft.   3. 12.56 sq. ft.
4. diameter: 1,900 ft.; circumference: 5,966 ft.; area: 2,833,850 sq. ft.
5. radius: .358 inches; diameter: .717 inches; area: .402 square inches

Quiz 38

2. 10; 5   3. five basketballs   4. 30 baseballs

Quiz 39

1. 42.43 yards   2. 3,140.06 ft.   3. 98.49 ft.   4. 8.25 meters   5. 509.12 ft.

Quiz 40

1.

	Mean	*Median*	*Mode*	*Range*
Week 1	$1\frac{1}{21}$	$\frac{13}{21}$	none	$1\frac{1}{3}$
Week 2	$\frac{16}{63}$	$\frac{1}{3}$	none	$\frac{3}{7}$
Week 3	$\frac{89}{126}$	$1\frac{1}{42}$	none	$1\frac{2}{21}$

Quiz 41

1. left: 88; right: 66; center: 66   2. 30%   3. 73–89

4. $\frac{75}{100}$, .75, $\frac{3}{4}$; other answers are possible.   5. 77%

6. A. $\frac{1}{8}$ B. $\frac{3}{8}$ C. $\frac{1}{4}$ D. $\frac{3}{4}$ E. $\frac{1}{64}$ F. $\frac{1}{56}$ 7. $\frac{16}{21}$ 8. .796   9. 83%   10. 0   11. 1

Quiz 42

1. Vargas: 71°; Carter: 68°; Martinez: 221°   2. Vargas: 316°; Carter: 44°; Martinez: 0°
3. 62.5%   4. 1.39%   5. 155°   6. rounding error

Quiz 43

1. stem-and-leaf plot:

0 | 0
1 | 2
4 | 1, 4, 5, 9
5 | 1, 6
7 | 1, 3
8 | 2, 5, 9

histogram:

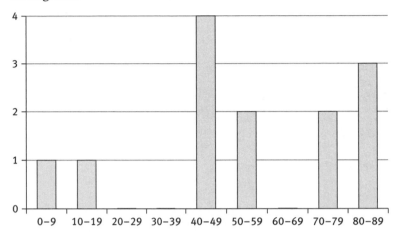

2. stem-and-leaf plot:

0 | 0, 0, 0, 4
1 | 1, 6
2 | 9
3 | 1, 7
4 | 7
5 | 7
7 | 1
8 | 4
9 | 6
10 | 1

histogram:

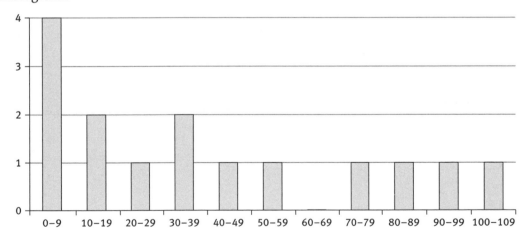

Quiz 44

1. Positive correlation is a statistical relationship in which an increase in the value of one variable is accompanied by an increase in the value of the other variable.

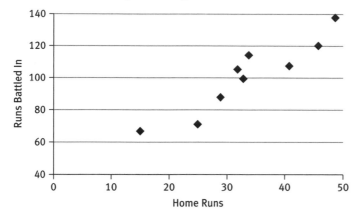

2. Positive correlation is a statistical relationship in which an increase in the value of one variable is accompanied by an increase in the value of the other variable.

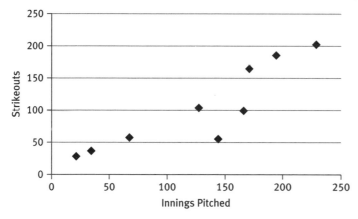

Quiz 45

1.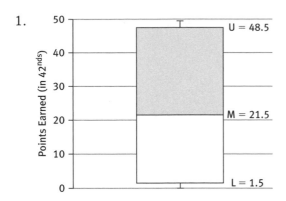

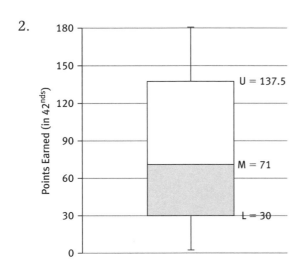

2.

Quiz 46

Answers will vary.

Pretest/Posttest

1. $3\frac{20}{21}$ 2. $\frac{1}{3}:\frac{1}{2}$ or 2:3   3. .857   4. $4\frac{17}{21}$ 5. $\frac{3}{4}$ 6. $2^2 + 3^2 + 5$ 7. $2\frac{1}{14}$ 8. 78 hits in 91 games

9. 81   10. range: $\frac{13}{21}$; mean: .494 or $\frac{20.75}{42}$; median: $\frac{17.5}{42}$; mode: none   11. A. $\frac{5}{48}$ B. $\frac{1}{4}$

12. 88%   13. $147,875   14. 319,680,000 square inches   15. 2,434.83 ft.   16. A. $\frac{3}{11}$ B. $\frac{3}{11}$
17. $R = 3$

Appendix: Lesson Plans

How to Begin the Game

Notes

You may wish to administer the Pretest (see Chapter Six) before the game begins.

The handouts in this text and in the accompanying student workbooks are the same.

Please do not despair if lesson numbers two or three do not go as smoothly as planned. When I first introduced the game to students, it wasn't until the third week that students began to master the computational process for their teams. If you can hang in there, your students will be rewarded with a rich learning experience full of excitement.

Lesson 1

Distribute Handouts 1 and 2. Review the handouts with students. (If students do not have student workbooks, I recommend that they keep all their Fantasy Baseball and Math materials together in a folder.) Have students select players.

Homework: Students complete their team roster, staying under the salary cap, and complete Handout 2.

Lesson 2

Distribute Handouts 3, 5, and 6. Have students practice computing points for players on the English Bulldogs, using the Bats-Rats box score.

Homework: Students complete Handout 6.

Lesson 3

Distribute and explain Handouts 4, 9, and 11. In addition, hand out box scores from one day in the previous week or let students access data online. If you hand out box scores from newspapers, you may need to include box scores from two different days to ensure that all teams are accounted for. Most teams play on Tuesdays and Wednesdays, but it's possible that one or more teams may not play on any given day. Have the students compute the points earned by their players for the first week on Handout 9. After the students have computed points, if time permits, have them trade papers with a peer and verify their peer's computations. If the computations are correct, have students sign their name at the top of their peer's worksheet. Students can then post their confirmed scores on the Fantasy Baseball bulletin board, using Handout 11.

How to Introduce Graphing Activities

Lesson 4

Show students how to construct circle, stacked-bar, or multiple-line graphs. (Note that circle graphs may be too advanced for some students.) For the stacked-bar and multiple-line graphs, particular attention needs to be given to the intervals on the y-axis (see the last paragraph in Chapter Three).

Each week, students compute points, update graphs, and complete worksheets that dovetail with the concepts they are studying in their textbook. As soon as students are comfortable with computing points (usually after a few weeks, depending on grade level), you can introduce the total points equations (Handouts 7, 8, and 10).

Tips and Suggestions

- There are several ways to verify the accuracy of students' work. As previously stated, students can sign their name at the top of a peer's worksheet to verify their work. Second, when students compute points, they can write the points earned by their players on the board. If several students post the same points earned for a player, they are probably on the right track. Students can use this information as a guide to check their work. However, students cannot simply copy the points from the board because they still have to show their work on their own worksheets.

- You can assess students via the pretest and posttest, forty-six quizzes in this text, or have students compute points for your team, using a scoring system you choose. Because you have your own fantasy team, you will already have computed points earned and thus will have the answers. Students can also construct one or more of the three types of graphs for your team for a given week.

- Students can also create their own scoring systems and use them to compute points. To do this, they will need to evaluate scoring systems that they have previously used, which gives them opportunities to use evaluation, the highest-level thinking skill, according to Bloom's Taxonomy.

- You can also play Fantasy Baseball and Mathematics in groups. Place students in triads to compute aggregate points. The different groups compete against each other. Thus, two games can run concurrently: one game for individuals and another game for groups.

Another alternative is that students can have more than one team. They can also play two sports at once. For instance, the schedule for professional football and baseball seasons overlaps in September, so students could play both fantasy sports concurrently.

- You can also play two games of Fantasy Baseball and Mathematics. In other words, halfway through the season, students can select new teams. This gives hope to students whose teams do not perform well during the first game. Students will be excited and highly motivated to select new teams.

- Dedicating a bulletin board to the game is a must. Students can post their teams, graphs, statistics, trades, and team rankings. You can include a "Wall of Fame," in which students post the most valuable players or the best bargains according to the salary cap. Conversely, you can also include a "Wall of Shame," in which students post the least productive players according to the salary cap.

- You can construct a large bulletin board in the shape of a baseball diamond. Each student is represented on the bulletin board by a baseball that circles around the bases from week-to-week as the season progresses. This is an effective way to visualize the rankings of the students' teams.

- You can link Fantasy Baseball and Mathematics to other disciplines. For example, you can post a map of the United States in which students can mark the cities that host professional baseball teams. You can give students blank maps in which they name the cities and states in which the baseball teams are located.

- Students can make their own spreadsheets to list their teams, weekly scores, and cumulative scores. They can also use computer programs to construct a variety of graphs.

- Each week students can be given an expository writing assignment in which they write about the math concepts they used and explain what they learned. A similar assignment can also be given at the end of the game.

- Students can also conduct research on the history of players. For example, they could write a report on their favorite player that included community service the player has been involved in, a history of the area where the player was born, or the college the player attended. Researching colleges that players attended may be the first exposure students get to higher education; perhaps this exposure can help students to set goals toward attending college.

- Prizes? You can take the top three and bottom three team owners to lunch at the end of the season, throw Fantasy Baseball pizza or ice cream parties, or inscribe the name of each year's winner on an inexpensive trophy, which would then stay at the school in perpetuity.

- Finally, you can make a transition to Fantasy Football and Mathematics at the end of the baseball season!